# SxI – Springer for Innovation / SxI – Springer per l'Innovazione

Volume 11

For further volumes:
http://www.springer.com/series/10062

# Sxl – Springer for Innovation / Sxl – Springer per l'Innovazione

Volume 11

Isabella M. Lami
Editor

# Analytical Decision-Making Methods for Evaluating Sustainable Transport in European Corridors

Springer

*Editor*
Isabella M. Lami
Regional and Urban Studies and Planning
Politecnico di Torino
Turin
Italy

ISSN 2239-2688 ISSN 2239-2696 (electronic)
ISBN 978-3-319-04785-0 ISBN 978-3-319-04786-7 (eBook)
DOI 10.1007/978-3-319-04786-7
Springer Cham Heidelberg New York Dordrecht London

Library of Congress Control Number: 2014943242



Printed on acid-free paper

Springer is part of Springer Science+Business Media (www.springer.com)

"But a commander in chief, especially at a difficult moment, has always before him not one proposal but dozens simultaneously. And all these proposals, based on strategics and tactics, contradict each other. A commander in chief's business, it would seem, is simply to choose one of these projects. But even that he cannot do. Events and time do not wait. […] An order must be given (him) at once, that instant. And the order to retreat carries us past the turn to the Kaluga road. And after the adjutant comes the commissary general asking where the stores are to be taken, and the chief of the hospitals asks where the wounded are to go, and a courier from Petersburg brings a letter from the sovereign which does not admit of the possibility of abandoning Moscow, and the commander in chief's rival, the man who is undermining him (and there are always not merely one but several such), presents a new project diametrically opposed to that of turning to the Kaluga road, and the commander in chief himself needs sleep and refreshment to maintain his energy and a respectable general who has been overlooked in the distribution of rewards comes to complain, and the inhabitants of the district pray to be defended, and an officer sent to inspect the locality comes in and gives a report quite contrary to what was said by the officer previously sent; and a spy, a prisoner, and a general who has been on reconnaissance, all describe the position of the enemy's army differently".

[Lev Tolstoj, *War and Peace*, The Cambridge World Classics, 2010]

# Foreword

## Infrastructure and Sustainable Growth: The European Perspective

Until the early 1980s the international geography of transport was dominated by the relationship between the world's major economic centres, namely the United States and Europe. This led to a rapid development of European ports positioned along the Atlantic coast, representing the points of easier access to the internal market. Over the years, this monopoly has increasingly consolidated, despite the construction of transhipment ports in the Mediterranean in the late 1990s. This allowed recovery of least part of the economies of scale and efficiencies of transport needed to bridge the competitive gap with the northern ports.

At present, the development of the economies of South-East Asia has completely changed the structure of trade routes giving a new focus to the Mediterranean (and therefore also Italy) ports. However, northern ports continue to increase their development not only because of the geographical advantages, tied to their proximity to major markets in Europe. They, therefore, have greater economic efficiency and reliability in the integrated sea–land supply chain, which still appears to compensate for increased travel times.

It is necessary to emphasise that the competition between the ports of northern Europe (namely: Rotterdam, Hamburg, Antwerp, Bremen, Le Havre and Zeebrugge) and the Mediterranean (Valencia, Barcelona, Marseille, Genoa, La Spezia and Piraeus) not only potentially concern the foreign, rival markets, but also the goods with origin/destination in the domestic market.

This competition is likely to increase in the years to come.

Until the last decade, the area served by a port has always been largely made up of contiguous territories.

In perspective, the increase in efficiency of long-distance connections, mainly due to the implementation of European corridors, will significantly change the face of this system. Ports orient their strategies to expand trade on expanding to more and more distant areas, despite the prevailing domain of other terminals. At the same time, the stronger economic and logistical areas of Europe feel the need to establish effective business relationships with the largest number of ports as possible, in order to have more options and greater flexibility in transport. Added to this are the growing markets of Eastern Europe, which have helped to shift the focus of international trade to the East and that today is a new and consistent basin to serve.

The economic crisis that has led to a sharp contraction of trade and which saw a time of uncertainty, has also led to an opportunity for the competitive repositioning of European ports. However, the economic situation is changing only slightly. After a disastrous 2009 (Rotterdam −9 %, from 10.8 million TEU in 2008 to 9.7 in 2009; Genoa −17 %, 1.8000000–1.5000000 TEUs—Data Port Authority), trafficking returned to growth in all ports, but with a reduction in growth rates (the amount of traffic of major European ports shows +10.26 % 2009–2010, +7.15 % 2010–2011, +3.32 % from 2011 to 2012).

The gap between the Northern Range ports and the Mediterranean is likely to increase, due to the current trend in the shipping market: it is the effect of "merging" (or the agreements between the major shipping lines), i.e. connections between the major shipping lines and port operators, and naval gigantism.

The future of the ports will also be dictated by the characteristics of the infrastructure that affect whether or not the access of new megaships and logistics services in the hinterland. Unlike the ports of the Northern Range, numerous Mediterranean ports do not allow the use of large vessels, either because of the characteristics of the infrastructure of the harbours (i.e. water depth, the size of the docks and availability of storage spaces) or because of the different economic dimension of the hinterland. Therefore, the Mediterranean ports do not allow exploitation of similar scale economies employed in Northern Europe.

The trend in data on the ships size is striking, as are the future predictions. Estimates of Ocean Shipping Consultants (OSC) which show that the number of container vessels with a capacity greater than 4,000 TEUs will be expected to grow progressively and will constitute 60 % of the world fleet in 2025. To a large extent this increase will contribute towards the growth of ships with a capacity of more than 10,000 TEUs.

The current monopoly of the Northern Europe ports in the international transport market leads to a noticeable impact on the environment both globally (i.e. more energy consumption and emissions due to longer journeys) and locally (i.e. the sensitive issue of noise pollution from passing trains in the Rhine valley of Germany). According to the conducted SoNoRa study in "New EU Freight Corridors in the area of the Central Europe" by the IUAV University of Venice (IUAV 2009) it is estimated that the forwarding of containers from the Suez Canal to Munich via the North Adriatic rather than by Rotterdam reduces $CO_2$ emissions of 135 kg per TEU.

In the three scenarios simulated by the OSC (natural evolution, recovery from the recession, prolonged recession), they assume an increase in the gap of container traffic handled by the ports of Northern Europe and the Mediterranean ports equal to 4–5 times that of 2010. If the estimates, at a global level, provided by the Ocean Shipping Consultants (OSC) are proved to be true, $CO_2$ emissions could still expand, leading to unsustainable environmental effects on European territory.

At a European level, faced with the alternative of either exploiting economies of scale and a further concentration of traffic on the ports of the North Sea (and the Baltic Sea) or bringing the Mediterranean port (and the Black Sea) to comparable levels of capacity and efficiency as of those of the North Sea, the latter option was favoured. This would allow the correct utilisation of a logistical transport asset which at present is not very sustainable.

This position is strengthened in the White Paper on Transport of 2011 "Roadmap to a Single European Transport Area—Towards a competitive and resource efficient transport", where it states that "in coastal regions there is a need for a greater number of efficient entry points to European markets to avoid unnecessary traffic flows across Europe. Seaports play an important role as logistics centres but require efficient connections with the hinterland" [1].

The White Paper on Transport updates the European Union's objectives in the field of transport until 2050. The Transport 2050 roadmap is intended to eliminate the major barriers and bottlenecks in many key areas of various sectors: transport infrastructure and investment, innovation and the internal market. The goal is to create a single European transport area and a fully integrated competitive transport network which connects the different modes of transportation and allows a profound change in the transportation of passengers and cargo.

At the infrastructure level, in addition to the application of new intelligent systems for security and information, stand three goals related to optimize the effectiveness of the multimodal logistic chains and to increase the use of more energy efficient modes of transport:

- "Complete the European high-speed rail network by 2050. Triple the existing high-speed rail network and maintain the dense railway network of all the Member States by 2030, because by 2050 the majority of medium-distance passenger transport should be by rail.
- The multimodal TEN-T 'core network' should be fully operational throughout the European Union by 2030. In 2050 a network of high quality and capacity should be connected with a variety of information services.
- Connect all core network airports to the rail network, preferably high-speed, by 2050. Ensure that all major seaports are sufficiently connected to the system of rail freight and, where possible, inland waterways" [1].

The strategies in Europe are not only oriented to infrastructure development but also to promote actions of management and regulation of the transport sector, for example, the creation of a competitive market, the harmonisation of rules, simplification of procedures, the use of new information technologies for the safety and efficiency of transport and the internalisation of external costs of transport.

By 2030, 30 % of the transportation of goods by road should transfer to other modes of transport such as rail or inland waterways: in 2050 this figure should rise to 50 %. There is a real need to promote more sustainable forms of transport other than road transport (i.e. rail, inland waterways and motorways of the sea), in order to reverse the trend of recent years, which has seen a steady increase in freight demand (if we exclude the crisis of 2008–2009) and a much higher road transport quota than in the past.

The situation at a national level, however, shows great heterogeneity. Some countries are more advanced than others in the development of integration between transport modes. These countries include Germany, Belgium and The Netherlands that use most of their inland waterways and Austria, Finland and Sweden, which have a well-developed rail network that is conducive to a more balanced modal split.

Italy starts from a worse situation than most European countries, although it has improved its performance since the crisis: 88 % of the value of goods by road in 2007 has risen to 91 % in 2009. Due to the decrease and instability of the quantity of transported goods, roads were more favoured as they were more flexible than railways but recovered to reach 86 % in 2012. Despite the slight improvement, Italy is still bringing up the rear in respect to the nations with which it was compared, and this demonstrates the need to act quickly and with adequate investment.

On 17 October 2013, the European Commission published the final draft of the TEN-T network after consultations between the Member States and stakeholders. The new maps show the nine main corridors, one less than proposed in 2011. This network will connect 38 major airports with rail connections that lead to the major cities (one more than the original 37) and will comprise 15,000 km of railway lines converted to high-speed and 35 cross-border projects meant to reduce bottlenecks. The nine main lines are divided into two north–south corridors (Scandinavian—Mediterranean and the North Sea–Mediterranean Sea), three east–west corridors (the Mediterranean, the Atlantic Ocean, the North Sea–Baltic Sea) and four diagonal corridors (Rhine–Danube, Rhine–Alps, the Baltic–Adriatic and Eastern Europe–Eastern Mediterranean).

The new transport core network will be complemented by an extensive network of lines (comprehensive network) that connect to the core network at regional and national levels. The ultimate goal of the new guidelines of the European Union is to ensure that gradually by 2050, the distance from the main network would be no further than 30 min for the vast majority of European citizens and businesses.

The nine corridors mark a huge advance in the planning of transport infrastructure. The experience gained already shows the considerable difficulty of making cross-border projects and other transportation projects in a coordinated way in different Member States. The plans and governance structures of the new corridors will greatly facilitate the implementation of the new core network.

The community funding necessary to achieve the objectives are set to triple over the period 2014–2020 up to 26 billion Euros, which will serve as the capital of "goodwill", in order to stimulate other investments in the Member States intended to complete difficult cross-border links and lines that would not otherwise be built. According to estimates, the cost of implementing the first phase of funding of the core network will amount to 250 billion in the period 2014–2020.

It transpires that European policies, such as the transport infrastructure policies of the TEN-T and the pattern of land use implicit in them, tend towards a rebalancing of the territorial organization of production and trade, which would bring the Mediterranean back to the centre of the logistic continental arena.

In this context, the degree of performance of the integrated systems—Ports—corridors—will play a key role on several levels: the infrastructure level, linking the availability and degree of congestion in the network of land transport connected to the inland: the level of transport services, the availability and effectiveness of liaison services, and, finally, the logistics level linking the efficiency of the transport chain and its integration into the overall supply chain.

"Intermodal" is the keyword for the development of a logistics system that is both efficient and sustainable, which respects the traversed territories and drawing its strength from the resources and the peculiarities that they themselves are able to generate and exploit. It is through the integration of infrastructure and territory that the efficiency of the logistics system can be appreciated as a whole.

To take this opportunity, however, it is necessary to promote clear national policies consistent with the European Union, to have the political courage to make choices and prioritise actions: policies that are able to specify requirements (demand analysis), to set goals through concrete numbers, programmes, actions and infrastructure with the consent of the population and to identify the necessary funding sources.

For these reasons it is essential to develop valuation models in order to quantify and qualify the effects (direct and indirect, positive and negative) of certain decisions of logistics and transport systems on the territory and the environment.

Never in these circumstances, as they have historically done in the field of transport of goods and people, do we need to exercise the maximum of predictive ability, to prevent, or at least reduce, the risk of an irreversible bad decision.

Hence, many evaluation tools are available in the decision process to support the transformation of the territory: multi-criteria techniques, analysis for the selection of alternative scenarios, economic analyses of the investment (such as cost-benefit analysis, opportunity costs). These are tools through which to operate in order to make an appropriate choice to resolve the complexity of the shared problems.

Turin, March 2014 Riccardo Roscelli

## References

1. European Commission (2011) White paper on transport of 2011—Roadmap to a single European transport area—towards a competitive and resource efficient transport. http://ec.europa.eu/transport/themes/strategies/2011_white_paper_en.htm
2. SoNorA (2009) New UE Freight Corridors in the area of the central Europe, final report edited by Research Unit "Transport, Territory and Logistics" (TTL)-University IUAV of Venice. https://www.port.venice.it/files/page/studiosonoraco2.pdf

# Acknowledgments

This book is the result of many challenging discussions between engaged scientists, local representatives, and experts from five countries in Europe. The framework that provides us this fruitful exchange has been the European Project CODE24 (2009–2014), which was approved under the Strategic Initiatives Framework of the INTERREG IVB NWE programme. The project has opened up new angles and assessments on similar problems in parallel circumstances. The research was a multi-national collaboration, and as such brought to bear institutional peculiarities and their importance in searching for solutions.

I would like to thank you all the Code24 partnership, also the people who didn't directly provided a chapter for this book, because they have all contributed to the excellent results achieved in the project and to our research work.

Turin, March 2014 Isabella M. Lami

# Contents

# Introduction

**Eurocorridors: A Political Project Developed Over Space and Time**

**Abstract** Eurocorridors are characterised by intensive transport flows and dynamic patterns urbanisation. They are also considered to be the backbone of powerful spatial and economic forces in the areas that connect urban regions. Eurocorridors are categorised by four parameters dimensions: axis of infrastructure, economic development, urbanization and institutional development. These four dimensions are viewed differently, both qualitatively and functionally, in that they may coexist, yet may act at quite different levels. Although the term "corridor" clearly suggests the concept of connection and access, it may fail to adequately represent all the aspects (subtle but crucially) related to the aforementioned dimensions. The problem of scale and scope suggests a preference to natural geographic shape, rather than an institutional structure, and an idea of homogeneity rather than distinctiveness. Although it is more desirable for infrastructure and institutional connections to perform effectively, it is uncertain how such high levels of consistency are "necessary" in terms of economic development or urbanization.

This chapter illustrates how the book addresses the four dimensions of the corridor, with the goal to elaborate, in both theoretical and empirical ways, the topic of analytical decision-making for evaluating sustainable transport with respect to investment in European Corridors.

## Transport Infrastructure and Territorial Transformations

The definition of infrastructure is often confused and shared with that of public works and socially fixed capital. The terms are sometimes used interchangeably (infrastructure as a work of public utility), or as terms to relate to each other, but with different meanings (physical work which requires a public initiative).

This is especially true for transport infrastructure. Traditionally, the market does not produce transport infrastructure for many reasons: because substantial investments are required (with a high capital/product ratio); because they are indivisible goods to be produced in their entirety (partial completion does not ensure results in the short term), because the rate of profit is relatively small and can often only be achieved over a long period of time. However, it is undoubtedly the ability to produce new social fixed capital, through general conditions that is useful for the welfare and development of the community.

In the White Book of 1993, infrastructure is broadly defined as "common capital stock" whose spatial distribution creates productivity differentials between regions and countries and affects production costs. The intrinsic features of the infrastructure are taken from the theory of public goods (immobility, indivisibility, inability to substitute) and show the same abstract versatility. In other words, infrastructure is defined as a capital asset, a factor of production, and at the same time as a public good, from which all can benefit [1].

Traditionally, the construction of a transport infrastructure is seen as a "source" of benefits for the area in which it is situated, guaranteeing competitiveness and economic growth. However, this tends to obscure the complex distribution of advantages and disadvantages, and the potential conflicts of social and territorial nature that the implementation of any "major work" determines [2].

In fact, if infrastructural interventions have a positive effect of increasing the chances of individuals and promoting the dynamics of development, it will not have the same effects on all the actors, at all levels and in all geographical areas involved in the process. Infrastructure upsets the status quo and brings out the typical forms of conflict that arise from the realisation of the major works of public interest: local versus global, diffuse interests versus concentrated interests; economic versus environmental issues [3]. It needs to be mentioned that territorial development is not a neutral process, as it can involve conflicting interests and strategies. The implementation of development policies can paradoxically generate further and different imbalances.

Since the dawn of human history, there has been infrastructure, and yet in the nineteenth and early twentieth centuries it has impacted the landscape more than the current realization of infrastructure does today. Just think for example of the great railway lines, Alpine crossings, ports, canals, works that belonged to the strategic visions of great connections through Europe and beyond. This idea that infrastructure is as a disruptive factor to the landscape is not such a current notion, but comes from a historical perspective. The difference is that perhaps nowadays infrastructure should be seen as integrating part of the landscape.

The infrastructural works are therefore a stimulus to launch processes of social and physical structuring of the local system, through the mobilization by the local actors of planning and organizational resources through their strategic visions and actions of governance. These visions and actions targeting are in fact a catalyst and an accelerator towards territorial transformations [4].

Major works are always territorial because they structure the area at their own level, whether being local, regional or national. Their nodes and their paths appear as opportunities and threats for the territories of local and regional level crossing or which are within the fields of externalities they generate or modify. Such works in fact, do not just exploit local situations, but can and should trigger cumulative processes of local development such as the dynamics in which the exploitable land resources are created through cooperative interactions between people. This surplus that local actors create, interacting with each other at the same time with their own milieu and supra territorial networks, is defined as "territorial added value" [5].

The reasoning with regard to the effects of transport infrastructure on the territory becomes more complicated and even more potentially disruptive when they are made at a European level, defined as European corridors or "Eurocorridors".

## European Corridors

In architecture the concept, characteristics and performance of a corridor are reasonably clear: its primary purpose is to give access to a variety of different rooms, areas or activities. To be functionally and economically viable, it is necessary to build a corridor as short as possible while still providing effective access to all the rooms. However, if the corridor is well made it should be a space with interesting architectural style. By contrast, corridors of development and infrastructure may need to perform in a variety of different roles, that could act in conflicting ways.

This concept of corridor is not easy to define, as it is not always clear which territories nor political parties are involved.

Normally, the corridor is seen as a "multifunctional backbone", including transport infrastructure for people and goods, of high-level services (research, logistics, etc.), creating spatial and environmental effects within a framework of cross-regional and local policies. Therefore, this very broad definition of "multifunctional backbone" covers a wide range of services to and from a specific territory, where the development of infrastructure is seen as a strategic driver for the transformation of the territory and not only as a project that meets the needs of a particular sector.

The corridor as an infrastructure axis is often defined in terms of traffic engineering. A corridor is used in this sense when developing or improving (interconnected) infrastructure modalities on a particular route. A simultaneous approach to the various modalities within a corridor offers important advantages, such as opportunities for bundling, and with it, a restriction of further criss-cross traffic. The corridor referring to the economic development axis supposes an implicit or explicit relationship between opportunities for economic development and major traffic axes. This point of view assumes that the spatial results of functional economic activities are strongly determined by the infrastructure network. The corridor pertaining to the urbanisation axis implies that the infrastructure network functions as the basis for the direction of future urbanisation for residential and work activity. All three interpretations of the corridor concept are a suitable example of what could be referred to as implicit theory: "the assumption is that traffic and infrastructure are not only derived from social and economic processes but to a high degree determine these functions as well.

Following this logic, corridors have a considerable impact on spatial developments and spatial patterns" [6].

Romein et al. [7] add a fourth dimension: besides the corridors as an axis of infrastructure, economic development and urbanization, they underline the role of institutional development. The latter illustrates the necessity of co-ordination during the decision-making process, the potential difficulty of integrating policy-making and management across various institutional boundaries, and the differences in operating standards, regulations and taxation regimes across the borders.

The four dimensions should be viewed as having differing qualitative and functional elements: the dimensions may coexist but they can also be seen as acting at quite different scales. Although the term "corridor" clearly suggests the concept of connection and access, may fail to adequately represent all aspects (subtle but crucial) related to the above four dimensions. There is also a problem of scale and scope that also suggests a natural geographic shape and linear rather than an institutional structure, and an idea of homogeneity rather than distinctiveness. Although it is more desirable for infrastructure and institutional connections to perform effectively, it is less clear how such high levels of consistency are "necessary" in terms of economic development or urbanization [8].

One of the main challenges in the spatial planning of Eurocorridors has been the need to engage in different types of collective action. This is because the development of corridors not only cuts across regional and national administrative boundaries, but also intersects with issues of economic development, environmental sustainability, and quality of life. These are complex issues requiring a collective approach to spatial planning that involves a broad network of actors across different sectors. Such an approach can be extremely challenging in practice, because the actors involved view spatial planning differently, due to their specific interests, perspectives and spatial scopes, which can lead to intractable levels of conflict. In summary, the spatial planning of Eurocorridors can be seen as a complex multi-scale and multi-actor problem [9]. Therefore effective development of Eurocorridors requires collective decision-making processes that include and involve the respective actors and the issues involved.

Hence the definition of the corridor is not a formalized term nor is it the result of a government's decision: it is one of the many policies where strategies, clusters and alliances between the various actors operate [10]. It can be said that Eurocorridors are a political project developed over space and time.

This book addresses all four dimensions of the corridor, with the main objective to elaborate on the topic of analytical decision-making, in both theoretical and empirical ways, for evaluating sustainable transport with respect to investment in European Corridors. Figure 1 illustrates how the theoretical contributions of this book can be characterised on the basis of this view. The case studies are (also) cross-compared to the four interpretations of the corridor.

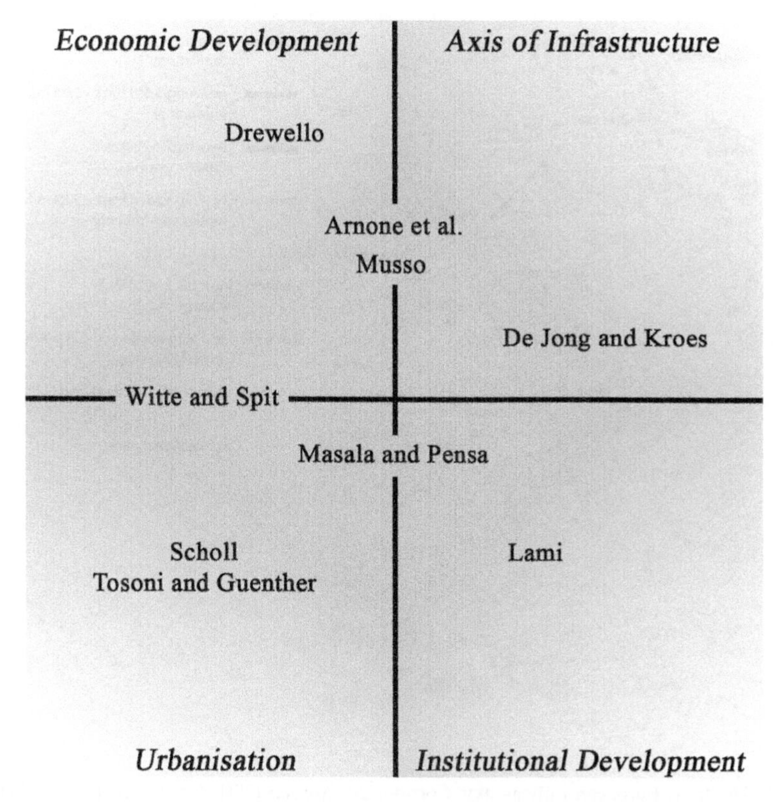


**Fig. 1** An overview of how the four dimensions of Eurocorridors are used in the theoretical contribution of the book

The starting point of the book was the work of collaborative and experimental research done during an Interreg IVB NEW Project, called "CODE24", regarding the corridor between Genoa-Rotterdam.

CODE24 aims at a coordinated transnational strategy to support the improvement and the development of the corridor. The overall objective is to accelerate and jointly develop the transport capacity of the entire corridor by ensuring optimal economic benefits and spatial integration while reducing negative impacts on the environment at a local and regional level. By focussing on regional aspects in the corridor area and joint development strategies, the project will strengthen the position of regional actors and stakeholders. It will provide planning tools and tailor made solutions to remove major bottlenecks and enable pro-active stakeholder participation. This encompasses both: the development of the railway system as well as a sustainable spatial development CODE24 involved 15 different partners from Italy, Switzerland, Germany, The Netherlands and France, including Cities, Associations of Regions, Port Authorities, Universities and Research Centres; collaborating with each other for 5 years to reach a common intervention

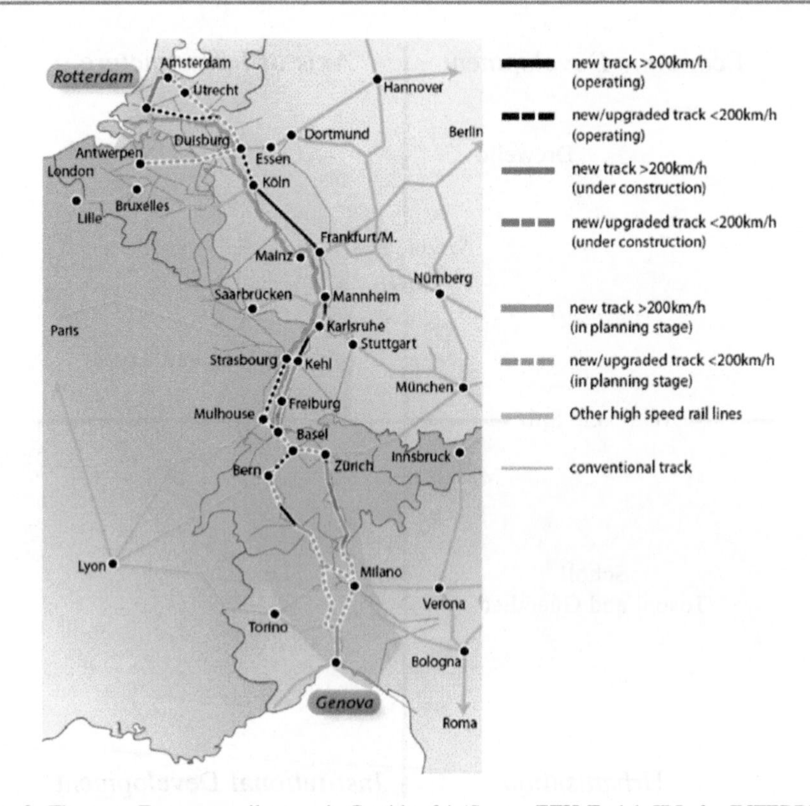


**Fig. 2** The trans-European railway axis Corridor 24 (*Source* ETH Zurich IRL for INTERREG IVB NWE project CODE24)

strategy. It was approved under the Strategic Initiatives Framework of the INTERREG IVB NWE programme and, as a result of the project, the European Grouping of Territorial Cooperation (EGTC) has been founded [11].

## The Corridor Genoa-Rotterdam

The trans-European railway axis (TEN-T) from Rotterdam to Genoa, initially defined with the number 24 (i.e. Corridor 24) and now, with the new grouping, defined as "Rhine–Alpine Corridors", intends to facilitate the interconnection of economic development, as well as spatial, transport and ecological planning. From a European point of view, there is a need to strengthen the links between countries to facilitate freight transport, especially considering the difficulties arising within a geographical context of orographic obstacles, administrative barriers and technical characteristics of the railway infrastructure which are often not compatible with each other. It constitutes one of the busiest freight routes of Europe, connecting the

North Sea ports of Rotterdam and Antwerp to the Mediterranean basin in Genoa, via Switzerland and some of the major economic centres in the western EU, having a catchment area comprised of 70 million inhabitants and operating 50 % (700 million tons/year) of the north–south rail freight (Fig. 2).

The European Union's objective is to double the capacity of rail transport on the axis by 2020, in order to encourage a modal shift of freight by rail: the main projects referring to this topic are the Swiss rail tunnel Loetschberg (opened in 2007), the Gotthard tunnel (the last barrier was torn down in October 2010 and the tunnel will be operational by 2017) and Mount Ceneri tunnel (which is expected in 2020).

Despite the importance of this connection, it faces infrastructural problems concerning freight traffic and passengers transport as many sections do not have adequate functional capacity in the corridor. Other problems are of managerial nature, due to the lack of coordination and interoperability at a trans-regional level caused by the presence of different transport services (freight, long distance, local traffic).

The economic and spatial development of the European axis is threatened to be limited by a number of major bottlenecks, a lack of trans-regional coordination, a decreasing consent among the involved population and the increasing difficulties due to uncertain financial resources.

Moreover, it is evident that the bottlenecks and main concerns within the corridor are predominantly caused by public acceptance, noise pollution, land management and landscape design, functionality of intermodal hubs and operational aspects as well as the management of planning processes and the financial issues related to major infrastructures.

Removing existing bottlenecks requires co-operation across political, organisational and technical bodies, focussing on four main topics:

(i) Railway and settlement development.
(ii) Environmental issues.
(iii) Integrated management of logistics.
(iv) Communication and involvement strategies for stakeholders.

In areas of conflicts regarding local planning competences and the supra-spatial planning issues, extensive forms of collaboration with the use of evaluation techniques able to support the decision-making process have to be explored and tested.

The construction of European transport infrastructures can be seen as a complex topic, where new values have to be taken into account. It is not a specific question of localisms, nor is it merely an issue of moving goods and people, nor does it affect only environmental, transport-related or town planning aspects. Tackling the issue of large-scale infrastructures involves dealing with a multitude of decision elements which require new trans-disciplinary approaches. Nowadays, the fundamental issue connected to large-scale infrastructures seems to be related to the definition of underlying agreements, rather than to the construction itself [12, 13].

In addition, the development from the perspective of the port areas (which are at the two ends of the corridor) would be to reduce the area pre-emptively as to minimize costs and the impact of transport (over land). The saturation of terrestrial links would otherwise prevent the growth of the ports risking the paralysis of the production structure, with huge economic costs of dispersion, worsening of environmental conditions and practically insurmountable constraints to the competition.

This is particularly important given the current trend of traffic flows through a few large-scale ports, where it is also possible to undertake programs for the expansion of terminals. Such growth is often accentuated by the continuous development of containerization which is particularly suited to maritime transport.

In general, recent innovations relating to port nodes are both capital and land intensive. When a new port is built but not designed or linked to the network of national transport properly, it presents higher costs for the local community. The local community is in fact called upon to deal with an increase in levels of congestion, atmospheric, noise and marine pollution, but also the removal of a public resource (the coastal areas) and an important asset (the space in general), becoming increasingly scarce which is subject to strong pressures for alternative uses.

In relation to the economic system of the surrounding area, the ports themselves may play a role in "structuring". This would create a series of effects, direct and indirect, to promote the development of multiple economic activities typically port transport and other activities, functionally linked thereto, the economies of spatial concentration could increase exponentially, if inserted in a framework of corridor's policies.

In the case of Corridor 24, which is characterized at both ends by international ports of very different sizes and characteristics, this needs to be stressed in a particular way.

## Outline of the Book

The main objective of the book is to elaborate, in both theoretical and empirical ways, the topic of analytical decision-making for evaluating sustainable transport with respect to investment in European Corridors.

Starting from the theoretical aspects of the Eurocorridors and various experience gained in the project CODE24, the book offers a series of tools for broad-spectrum evaluation designed to support decision making.

The book is divided into two main sections:

I. The first part is theoretical and provides an overview of relevant concepts in the field of transport issues; integrated strategies, governance and decision-making processes; assessment models and evaluation frameworks.

II. The second part is devoted to case studies, where three detailed applications of assessment models are applied to different areas of the Rotterdam-Genoa corridor.

I. In the Part I of the theoretical contribution, the main transport issues are analysed. Chapter 1 (Arnone et al.) illustrates how over the past 10–15 years the growth of car mobility slowed down in several European economies and, in some others stopped or turned negative, which may only partly be explained by the economic recession. To understand whether this phenomenon is transitory or permanent a combination of factors that affect mobility needs and patterns (mobility drivers) have been investigated. Chapter 2 (de Jong and Kroes) summarises some of the main issues in the analysis of freight transport. It starts by underlining the importance of studying freight transport, and which viewpoints can be taken. This is followed by a discussion of the different types of agents which are involved and their range of choices. The different modes of freight transport are compared, in particular road and rail transport, but also the other modes including water-based transport, air freight, pipelines and container based transport. The Chap. 3 (Musso) explains how the container is one of the many standardisation processes that make the industrialisation of maritime transport of general cargo possible but, at the same time, how the need for standardisation inevitably clashes with the need for change dictated by technological evolution. The conflict between evolution and standardisation, somewhat common to all industries, is particularly accentuated in the field of global container transport and is conditioning its explosive development.

The Part II of the theoretical contribution is more focused on the link between territory and transport. The starting point of Chap. 4 (Witte and Spit) is how the existence of bottlenecks in the European transport network is a persistent issue in European (spatial) policy. It also shows that the transnational transport corridor scale (macro), as well as the urban region and the local transport node scale (micro), are of importance. Therefore this chapter illustrates a methodology that is suitable for a simultaneous analysis on both the macro and micro level, since neither of the two levels is able to capture the full complexity of the transport bottlenecks occurring on transport corridors. Chapter 5 (Scholl) argues that the clarity in the joint infrastructure concept is not only essential for transport development, it is also important for the development and operation of a future network of cities and sites and thus for spatial development. Therefore, developing a joint infrastructure concept should be considered an integral task of European importance. This increases the importance of the supra-urban aspect of spatial planning, namely, regional planning as the middle level between national and community planning levels. The starting point of Chap. 6 (Lami) is the finding that, in contemporary society, territorial land-use conflicts have become more frequent and widespread and often lead to increased social conflicts. The conflicts concerning the location of so-called "undesirable facilities" such as "invasive" products (motorways, high speed train lines, waste disposal plants, etc.) are characterized by the protest of local communities fighting for the defence of their land from external intrusions. The chapter shows how, in this complex panorama, Multiple Criteria Decision Analyses (MCDA) can provide a very useful support. Chapter 7 (Drewello) focuses on the economic impact of regional accessibility and in different economic sectors in order to prepare a methodology of planning for efficient transport infrastructure investment along the corridor of Rotterdam-Genoa. The approach takes into consideration that the output of economic sectors depends differently on logistic services.

The Part III of the theoretical contribution concerns different models, tools and frameworks able to help the assessment procedure in the context of Eurocorridors. Chapter 8 (de Jong and Kroes) gives an overview of discrete choice analysis techniques, because it plays a relatively important role in transport analysis. Different types of discrete choice models are discussed: the workhorse of discrete choice modelling—the multinomial logit model (MNL), the nested logit and other Generalised Extreme Value (GEV) models, the probit model, the mixed logit and latent class models, ordered response models and aggregate logit models. Chapter 9 (Tosoni and Guenther) describes the collaborative assessment procedure, which is designed as a peer-learning activity, where participants from different backgrounds are asked to actively contribute to a process of knowledge formation, by sharing and consolidating. The procedure was developed in the framework of the CODE24 Interreg Project as a method to enable joint design of a shared spatial development strategy by the different stake-holders taking part in the initiative. The scale of the project area (about 1000 km long railway corridor) and the number of actors involved presented both a design and management challenge that the procedure tried to overcome. The Chap. 10 of the theoretical contribution (Masala and Pensa) proposes the use of the Interactive Visualisation Tool (InViTo) as a method for sharing information by using spatial data visualisation, also known as geo-visualisation, to support spatial decision-making and planning. InViTo is based on the idea that interacting with data can improve the knowledge process of users, while visualisation should contribute to increasing intuitive perception. The visual system works both 2- and 3-dimensionally to meet and improve users' skills in interpreting images.

II. The last part of case studies includes two applications of an integrated approach, combining Analytic Network Processes, InViTO and Strategic Assessment, for supporting the evaluation of alternative development strategies at two different scales, the metropolitan area (Chap. 11, Abastante et al.) and the European scale (Chap. 12, Abastante et al.); and a conjoint analysis exercise (Chap. 13, Arnone et al.), with the objective to provide a quantitative assessment of relevant quality attributes of freight transport services by estimating parameters based on empirical research into shippers' evaluation of freight services.

Turin, March 2014 Isabella M. Lami

## References

1. European Commission (1993) Growth, competitiveness, employment: the challenges and ways forward into the 21st century. White Paper http://ec.europa.eu/white-papers/
2. Dematteis G (2003) The development of territorial and network systems. In: Camagni R, Tarroja E (eds) Estrategies territorials. CIMBP, Barcelona
3. Bobbio L (ed) (2013) La qualità della deliberazione. Processi dialogici tra cittadini. Carocci, Roma
4. Clementi A (2007) Landscape and Urbanism in Italy. In: AA.VV, Landscape and society. Council of Europe Publishing, Strasbourg
5. Dematteis G, Governa F (2001) Urban form and governance: the new multi-centred urban pattern. In: Change and stability in urban Europe: form, quality and governance. Ashgate, Burlington
6. Priemus H, Zonneveld W (2003) What are corridors and what are the issues? Introduction to special issue: the governance of corridors. Transp Geogr 11:167–177
7. Romein A, Trip JJ, de Vries J (2000) Theoretical framework, report within the framework of action 3 of CORRIDESIGN. OTB Research Institute, Delft University of Technology, Delft
8. Chapman D, Pratt D, Larkham P, Dickins I (2003) Concepts and definitions of corridors: evidence from England's Midlands. Transp Geogr 11:179–191
9. Romein A, Trip JJ, de Vries J (2003) The multi-scalar complexity of infrastructure planning: evidence from the Dutch–Flemish megacorridor. Transp Geogr 11:205–213
10. Fubini A (2008) Corridor and territorial policies. In: Fubini A (ed) Corridor policies and territorial development: main infrastructure and urban nodes within Corridor V. Franco Angeli, Milano
11. http://www.code-24.eu
12. Bertolini L (2005) Cities and transport: exploring the need for new planning approaches. In: Albrechts L, Mandelbaum S (eds) The network society. A new context for planning. Routledge, London
13. Lami IM, Staffelbach L (2008) Evaluating the urban redevelopment of railway sites: Zurich Central Station and surrounding areas. In: Third Kuhmo-Nectar conference, transport and urban economics, 3–4 July 2008, Amsterdam

# Part I
# Theoretical Contribution: Transport Main Issues

# Passenger Mobility: New Trends

# 1

Maurizio Arnone, Tiziana Delmastro and Letizia Saporito

**Abstract**

This chapter analyses the main changes occurred to passenger mobility needs and patterns in the last decades and investigates which are the main demographic, economic, social and technological factors that have influenced and are influencing such changes (mobility drivers). Recent trends of the identified drivers have been analysed to draw possible development scenarios of passenger demand, forecast some likely effects on car mobility during the next 10 years and understand which are the main knowledge gaps and open issues that still need to be investigated.

## 1.1 Introduction

In past decades, the demand for passenger travel has developed roughly in line with GDP per capita and population growth, but there are strong signs that this close connection is weakening in advanced economies.

M. Arnone · T. Delmastro (✉) · L. Saporito
SiTI—Higher Institute on Territorial Systems for Innovation,
via Pier Carlo Boggio 61, 10138 Turin, Italy
e-mail: tiziana.delmastro@siti.polito.it

M. Arnone
e-mail: maurizio.arnone@siti.polito.it

L. Saporito
e-mail: letizia.saporito@siti.polito.it

I. M. Lami (ed.), *Analytical Decision-Making Methods for Evaluating Sustainable Transport in European Corridors*, Sxi 11, DOI: 10.1007/978-3-319-04786-7_1,

Passenger mobility patterns have shown big changes in recent years: the total number of vehicle trips as well as vehicle travel per capita show a declining or almost steady trend in many European countries. Car ownership is no longer growing, while demand for alternative transport modes is increasing.

The recession of 2008 might have played a role in accelerating mobility changes due to the increasing unemployment rate, the decreasing incomes and the rising fuel prices. However, looking further back at the last decades, we may notice that vehicle travel trends grew steadily only until 1995–2000, when they already started to level off and decline, despite continued population and economic growth.

This fact might suggest that the causes of the ongoing mobility changes are connected not only to temporary factors but also to other structural phenomena such as: an aging population, population distribution in urban and suburban areas, improvements to alternative transport services, the increasing concern for the environment and changes in users' habits and preferences.

This chapter, starting from an analysis of the main changes which occurred to the passenger mobility trends (see paragraph 2), identifies the most important mobility drivers, discusses how the changes of these factors affect travel demand and draws a number of conclusions regarding car mobility in the years to come (see paragraph 3). The final paragraph illustrates various open issues and discusses how transport modelling and planning should be improved to cope with passenger mobility changes.

## 1.2 Changes in Passenger Mobility Trends

There have been numerous studies on the effects the global recession has had on passenger mobility in recent years [1, 13, 17]. The most evident change is the stop of the steady growth of mobility observed in Europe after 2007 (Fig. 1.1). The dashed lines in the graph present two different slopes of the mobility curve and show how the average annual growth rate decreased from 1.6 % (1995–2007) to 0.2 % (2007–2011). Moreover, the provisional data collected for 2012 and 2013 confirm the same trend (UK: [7], Italy: [2, 3], France: [4]).

However, if these data are disaggregated by means of transport the effect of the crisis becomes less evident (Fig. 1.2).

Collective and air transport present steady positive trends, excluding some cyclic fluctuations, although air transport has grown more during the past 15 years, mainly due to globalization and low-cost flights [16].

Also car mobility does not present an obvious change after 2007. The graph shows a smoother deceleration in the observed period of time (roughly represented by the dashed line in the figure) which does not appear to correlate with the recession.

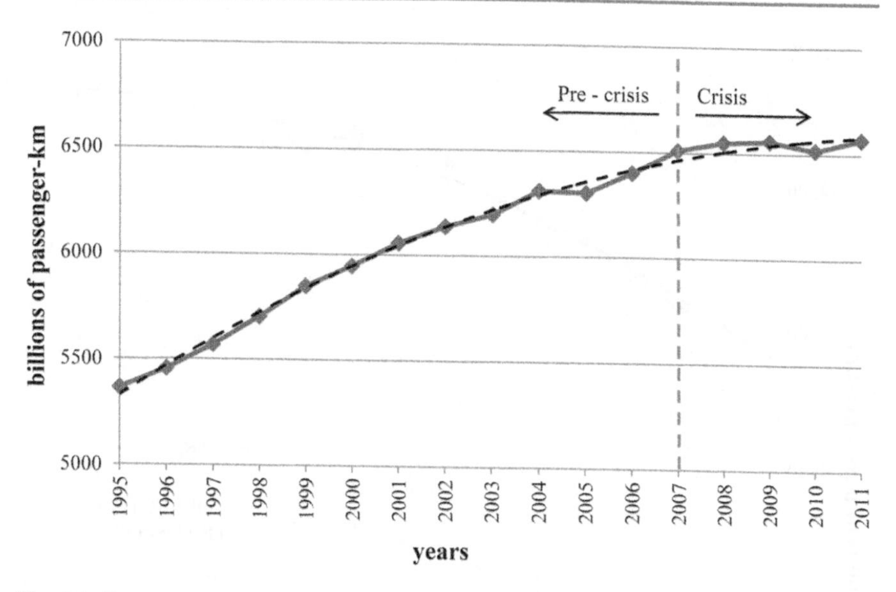


**Fig. 1.1** Passenger mobility in EU-27 (Eurostat data elaborated by the Authors [8–10])

In order to verify whether the decline of car travel started independently of the economic crisis, the car mobility data have been disaggregated per country and observed for a wider period of time (1970–2011).

We analysed eight European countries belonging to four different economic clusters, based on the GDP per capita and the time to recover from the continuing crisis. Data used to define the clusters are presented in Fig. 1.3:

- High GDP per capita economies and fast recovery (Germany).
- High GDP per capita economies and slow recovery (France, The Netherlands, UK).
- Medium GDP per capita economies and slow recovery (Italy, Spain).
- Emerging European economies (Poland, Bulgaria).

Data are illustrated in Fig. 1.4. All countries with high-GDP (France, the Netherlands and UK) and medium-GDP (Italy and Spain) with a slow recovery, reveal the influence the recession has had on the mobility, while countries with a fast recovery (Germany) seem not to be affected. However, the mobility trends of countries more affected by the crisis are very diverse: Italy shows an evident drop whereas UK and France present a gradual deceleration. This is because the economic factors (such as the increasing unemployment rate, the decreasing incomes and the rising fuel prices) are not the only ones that affect mobility. Socio-demographic and cultural factors are equally important and transport and land use policies also play a crucial role (see paragraph 3).

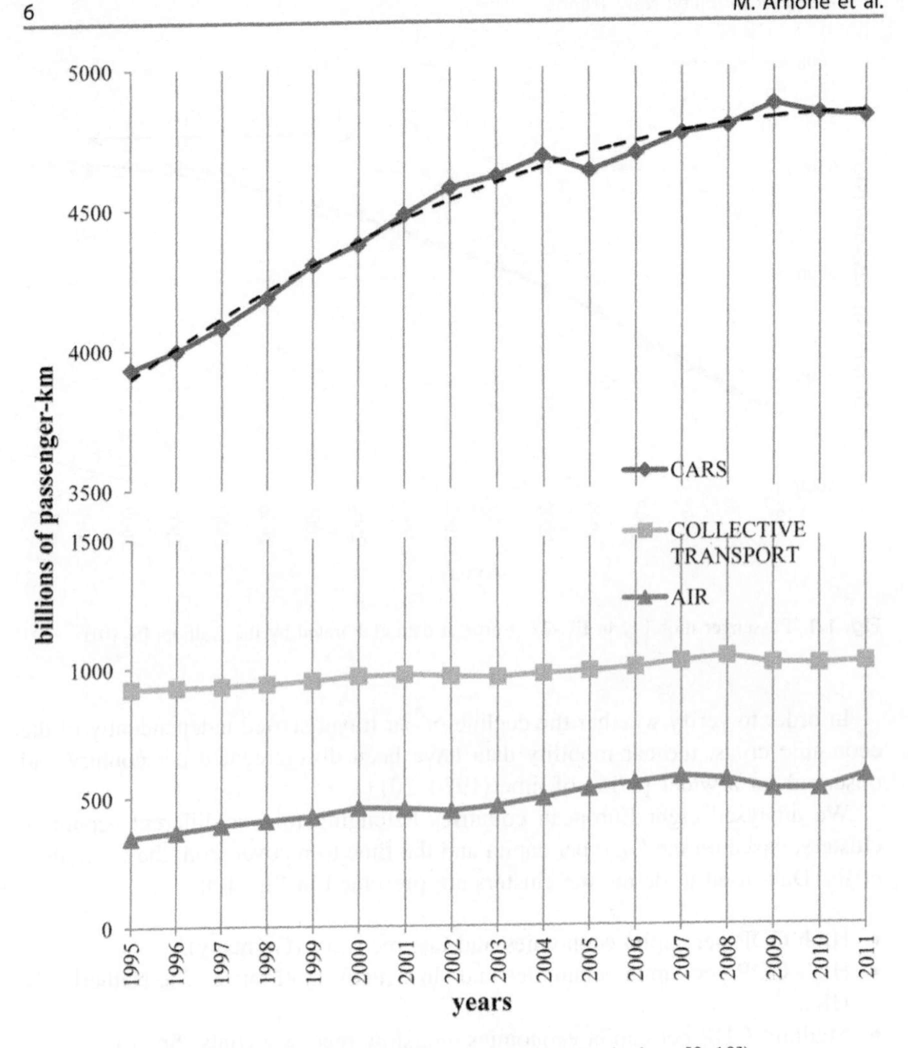


**Fig. 1.2** Modal split in EU-27 (Eurostat data elaborated by the Authors [8–10])

Moreover the illustrated data raise another interesting point. If we look at a wider period of time, the car mobility growth seems to have slowed down in both high and medium GDP economies, in several countries this process started earlier: Germany in approximately 1995 whereas the UK and The Netherlands at the beginning of the 1990s. Yet other nations present this trend only in the last decade: Italy from 2000 and France from 2005.

Spain has a different trend because, during the last 20 years, big investments in transport have been made to narrow the infrastructural gap between the other advanced European economies [11]. However, also Spain presents a slight stagnation from 2005.

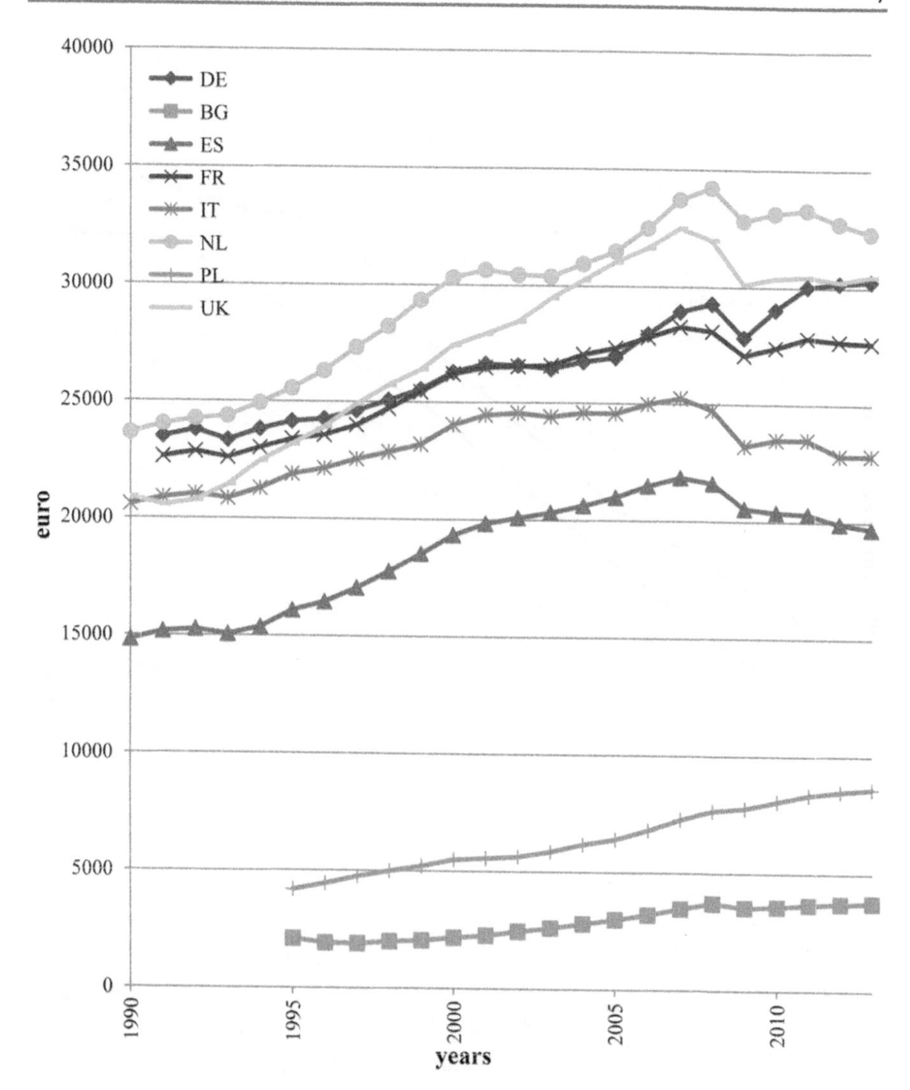


**Fig. 1.3** GDP per capita in a number of European countries (Eurostat data elaborated by the Authors [8–10]—calculated as millions of euro, chain-linked volumes, reference year 2005/N. inhabitants)

Emerging economies (Poland and Bulgaria) show a trend similar to Spain but are continuing to grow.

To resume, in advanced European economies car mobility slowed down during the last 10–20 years, after having grown steadily before, whereas emerging economies seem to still be in the car mobility boom period that other countries have already passed.

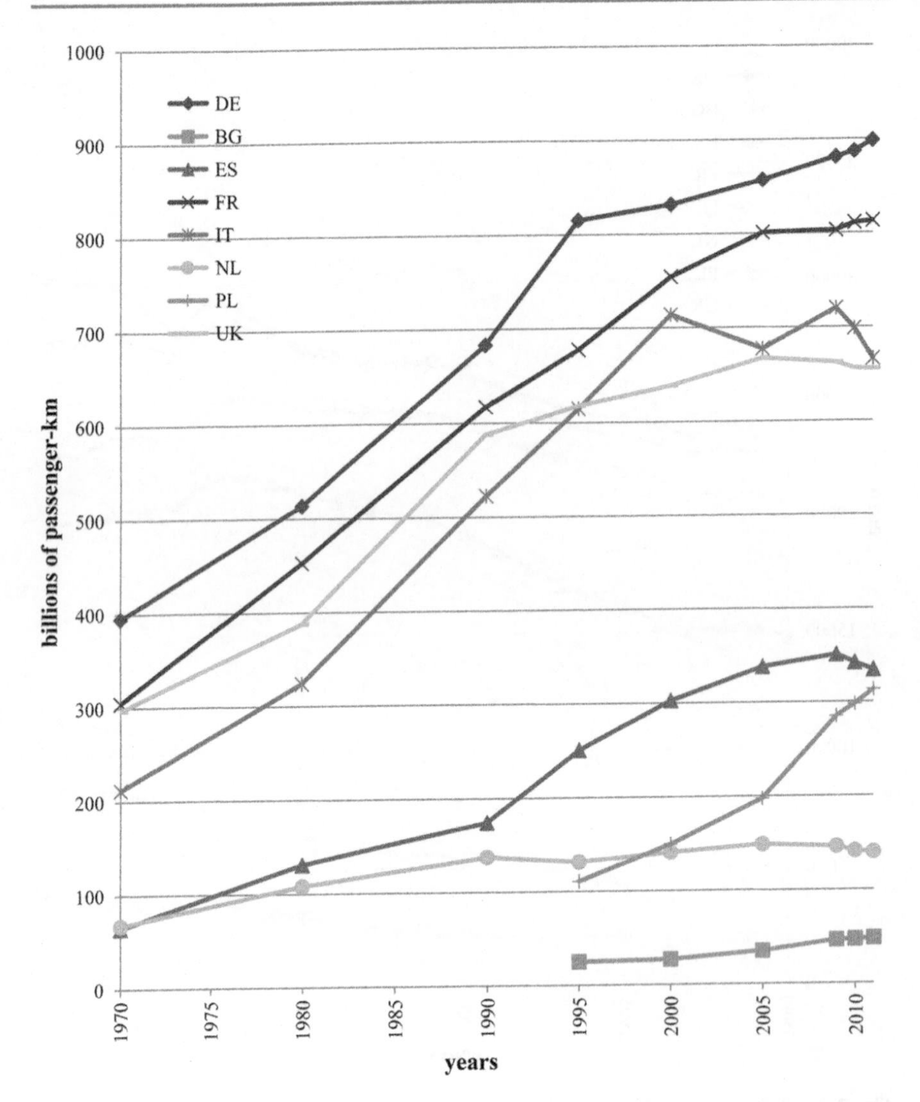


**Fig. 1.4** Car mobility in a number of European countries (Eurostat data elaborated by the Authors [8–10])

These considerations seem to be confirmed by the trend of car ownership per inhabitant (Fig. 1.5). Data indicate that car ownership growth rates started to decline after 1990 in most wealthy countries and appear likely to level off at a value of 0.5 car per inhabitant, if we exclude Italy (0.6) that has been suffering from car-oriented transport policies for several years [14].

In order to understand the possible causes of the aforementioned phenomena, the next paragraph identifies and analyses the main factors that affect mobility needs and patterns (mobility drivers).

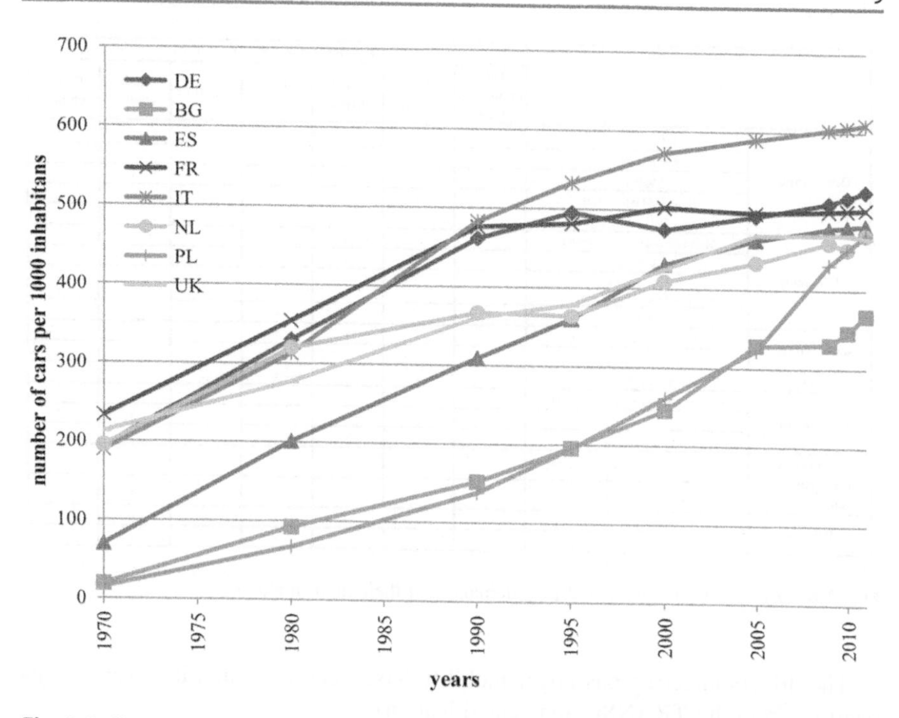


**Fig. 1.5** Car ownership in some European countries (Eurostat data elaborated by the Authors [8–10])

## 1.3 Factors Affecting Mobility

The main transport drivers affecting passenger mobility demand have been derived from previous studies and analysis conducted at European level, in particular the TRANSvisions study [16] and the "EU transport GHG: Routes to 2050?" project [15].

The selected drivers are listed in Fig. 1.6 where their main areas of impact are also indicated. Similarly to drivers, the impact areas have also been identified integrating the main outcomes of the analysed studies with transport experts' opinions.

In particular, seven impact categories were used to describe how drivers can affect passenger transport demand:

- Total number of trips: the overall transport volume in number of trips.
- Trip frequency: the number of trips per person or per household.
- Trip purpose: the distribution of trips by purpose.
- Trip distance: the length of trips in kilometres.
- Trip time: the time of day when the trips start.
- Mode choice: the choice of transport modality per trip.
- Route choice: the choice of the route if it is possible (only valid for road transport).

| Drivers | | Impacts on passenger mobility | | | | | | |
|---|---|---|---|---|---|---|---|---|
| | | Total number of trips | Trip frequency | Trip purpose | Trip distance | Trip time | Mode choice | Route choice |
| Demographic | Population size | x | | | | | | |
| | Age distribution | | x | x | | | x | |
| | Migration | x | | x | | | x | |
| | Household type & size | | x | x | | x | x | |
| | Re-urbanisation/urban sprawl | | x | | x | x | x | |
| Economic | Macro economic development | | x | x | | | | |
| | Income | | x | x | x | | x | |
| | Employment (rate, new types of employment) | x | x | | x | x | | |
| | Globalisation | | | | x | | x | |
| Socio - cultural | Time use/Leisure/Tourism | | x | x | x | x | x | |
| | Sustainable lifestyle | | | | x | | x | |
| Energy | Fuel availability and price | | x | | x | | x | |
| Technologies | New and improved transport modes | | | | x | | x | |
| | ITS | | | | x | | x | |
| | E - everything | | x | x | | | | |
| Transport infrastructures and services | Accessibility/Level of service/Safety | | x | | x | | x | x |
| | Integrated transport systems | | | | | | x | |
| Policies | Transport policies | | x | | x | x | x | x |
| | Land use policies | | | | x | x | x | |

**Fig. 1.6** Drivers of passenger mobility demand and their main impacts

The drivers affecting passenger mobility have been classified in seven groups (starting from the TRANSvisions classification):

1. Demographic drivers: including population size, age distribution, migration and household type and size. The increasing population size entails an increasing mobility demand, but recent trends show that also population composition is changing and needs to be taken into account. The proportion of elderly people in Europe is already higher than that of children: above a certain age, people tend to travel less, moreover they travel for different purposes than young persons (e.g. leisure trips). This may result in an increase of demand for collective modes of transport, even if nowadays elderly people represent a larger share in driving population than in the past. On the other hand, the decreasing number of persons in the productive age category and their progressive substitution by immigrants, filling employment gaps especially in low skilled jobs, is affecting mobility patterns in terms of number and purpose of trips and choice of cheaper transport modes. The decreasing tendency of household size (single or two-person households are more and more diffused, thus increasing the overall number of households) is also concerning transport demand, leading to an increase in car ownership per capita, due to the reduced opportunity to share the same car among household members, and a higher number of car trips per person, since single people are more willing to travel for non-systematic purposes. Single or two-person households have a propensity to live in urban centres: the growing trend of people to reside in urban areas or in the suburbs and the resulting urban sprawl (a shift in the location of activities towards the peripheries of urban agglomerations) influences passenger mobility. People

living in suburban areas mostly use cars or comodality and are willing to accept longer trip distances and earlier departures to reach their destinations (particularly commuters). On the other hand, in urban areas public transport services are more efficient and more frequently used and shorter trips are often carried out by walking or bike.

2. Economic drivers: including macro-economic development, income, employment and globalization. Economic development generates new activities and opportunities for a better and wealthier lifestyle which lead people to move with more frequency and for many different purposes. Similarly, a higher disposable income may induce an increase in car ownership and trip frequency as well as in long distance trips for leisure purposes. Even if, thanks to increasing wealth and revenues, the number of trips carried out for different purposes is constantly increasing, commuting still represents an important share (almost 30–35 %) of passenger transport demand [5–7, 12] and employment is an important mobility driver. The rising unemployment rate, generated by the economic downturn in the last years, saw the number of commuting trips decreasing; furthermore, due to the few job opportunities available, people are nowadays willing to accept an employment further away from their place of residence, thus affecting trip distances and trip times. Also innovative types of employments and contracts could play an important role in the change of commuters' mobility trends: the diffusion of part-time jobs will impact on the time of the day when people travel and teleworking will reduce the number of commuting trips. However there is still no large evidence of such changes and the debate concentrates on how the decreasing number of commuting trips due to home-based work, could be substituted by other types of trips: several studies demonstrated that people are disposed to spend a certain amount of their personal income (about 10–15 %) and of their time (about 1 h per day) on travel [16]. Finally globalization and the widening of European borders are inducing an increasing number of long distance trips by new low-cost transport modes (especially air travel).
3. Socio-cultural drivers: the improvement of living standards has led to an increase in leisure time that people can spend at home or outside, thus rising transport demand and trip frequency at different times of the day. Due to the availability of spare-time and to new travel opportunities and cheaper connections throughout Europe, tourism has also developed, leading to a shift of mobility demand to fast transport modes (particularly air transport). Another increasing trend affecting mobility patterns is the rising consciousness of sustainability: in transport this translates into an increased number of trips by means of walking, cycling and using public transport, which may also lead to reduce the trip length. Moreover, the car is not seen as a status symbol anymore nor the only way to provide freedom of movement, this view is predominant among young people.
4. Energy: currently fuel availability and price represent an important driver for passenger demand. Oil prices generally have a limited effect on traffic demand since their rise is usually balanced by the increase of personal income and improvements of vehicle fuel efficiency. However, as a consequence of the

economic crisis and of the ever increasing oil prices, the price elasticity of demand with respect to fuel costs has increased, meaning that transport sensitivity to energy costs is higher than before and this resulted above all in road traffic reduction.

5. Technologies: technology developments at different levels are influencing mobility patterns. New developing fuel efficient engines will allow to improve many means of transport (cars, buses, airplanes) and to reduce trip costs thus enhancing the use of alternative modes and increasing travelling distances. Through Intelligent Transport Systems diffusion (e.g. variable message signs, in-vehicle technology, route planners, etc.) trip quality is continuously being improved and, thanks to the opportunity of gathering, sharing and disseminating real time information to travellers, the management and organization of trips, particularly by use of collective transport modes, has become easier. On the other hand, as a result of the continuous improvement in digital communication, an increasing number of activities can now be done through the Internet, without the need of moving (E-Everything). E-shopping, e- conferencing, e-education, e-banking, e-entertainment seem to lead to a strong decrease of physical mobility since real trips are substituted by virtual ones. For example, on-site meetings can be often substituted by online meetings, saving time and money on distant trips; banks and universities can do much of their business online; furthermore nowadays an ever-increasing share of leisure time is being spent on the Internet. However, also trip generation effects occur: in addition to business-to-consumer (b2c) e-commerce, many consumers are frequently using the Internet also to buy from or to sell to other consumers (consumer-to-consumer e-commerce, c2c). Weltevreden and Rotem-Mindali [18] found that, in the Netherlands where c2c market overweighs b2c in terms of sales volumes, b2c e-commerce led to a net reduction in the number of shopping trips and distance travelled by consumers while c2c e-commerce generated the opposite effect. Moreover the widespread use of social networks, allowing people to keep in contact even if far away, and the easy access to all kinds of information (events, advertising, last-minute deals, etc.) may encourage the number of trips per person and the travelled distance to rise.
6. Transport infrastructures and services: the development of new infrastructures and transport services and particularly the improvement of their level of service, accessibility and safety aspects, creates new and better opportunities for mobility which boost the trip frequency and distance and impact on route (for car trips) and mode choice. The choice of alternative transport modes and a reduction of car trips can be strongly affected by efficient integrated transport systems which give people the opportunity to travel for long distances without long waiting times for transhipments and streamline passenger flows.
7. Policies: transport and land use policies at European, national and local level are very important driving factors affecting transport since they have the possibility to change the mobility drivers described above, acting, for example, on traffic and demand management, on transport planning or on urban redevelopment.

**Table 1.1** Likely effects of the main drivers on car mobility in the Piedmont Region in the next 10 years

| Drivers | | Effect on car mobility |
|---|---|---|
| Demographic | Population size | ↗ |
| | Age distribution | ↘ |
| | Migration | ↗ |
| | Household type and size | ↔ |
| | Re-urbanisation/urban sprawl | ↘ |
| Economic | Macro-economic development | ↔ |
| | Income | ↔ |
| | Employment | ↘ |
| | Globalisation | ↔ |
| Socio-cultural | Time use/leisure/tourism | ↔ |
| | Sustainable lifestyle | ↘ |
| Energy | Fuel availability and price | ↘ |
| Technologies | New and improved transport modes | ↘ |
| | ITS | ↔ |
| | E-everything | na |
| Transport infrastructures and services | Accessibility/level of service/safety | ↗ |
| | Integrated transport systems | ↘ |
| Policies | Transport policies | ↘ |
| | Land use policies | ↘ |

The recent trends of the aforementioned drivers have been investigated focussing on the area of the Piedmont Region (located in north-west Italy): data coming from different official sources, such as the national and regional statistical databases or regional mobility surveys, were collected and analysed. Through desk research integrated with data analysis and with the opinions of experts in several different research fields (transport, land planning, urban development, tourism, social sciences) development scenarios of such drivers and some likely effects on car mobility during the next 10 years were forecasted (Table 1.1). The reported results come from a preliminary phase of the analysis that still needs to be extended with focus groups and questionnaires especially aimed at filling the knowledge gaps about emerging drivers.

Even if it is not possible to predict future travel demand with precision, the main factors analysed may reveal that in the near future an increasing portion of travellers will prefer to drive less and rely more on alternative modes, if they will be comfortable, convenient and affordable. However, cars will continue to be an important means of transport.

## 1.4 Conclusions

Over the past 10–15 years, the growth of car mobility slowed down in several European economies and, stopped or turned negative in others.

The recession may explain only part of this change. To understand whether this phenomenon is transitory or permanent a combination of factors that affect mobility needs and patterns (mobility drivers) have been investigated.

We started from the main findings in literature and updated or integrated them with other driving factors gathered from the analysis of the Italian context.

Then, as an example, the drivers have been investigated in order to present their likely effects on car mobility in the Piedmont Region (in north-west Italy) in the next 10 years. The main results of this analysis confirm the findings of similar studies [13] and show an expected decrease of car use in the future (although it will continue to be one of the main means of transport), in favour of other more sustainable modes, such as collective transport, bicycles and walking.

Nevertheless some open issues still need to be considered.

Firstly, uncertainty on the development of some crucial drivers remains.

The most important doubts are related to the economic crisis. Is it a temporary cyclic crisis or does it hide a more structural change of our economy and society? Will we return to the pre-crisis levels of growth and, in that case, will the correlation with the mobility be the same? Or will new economic patterns and new typologies of employment, together with new welfare systems, change this relation?

Secondly, while some explanatory factors are fairly well understood, others remain unclear.

The contribution of several drivers is almost unknown, usually because of lack of data. For example the wide success of the new information and communication technologies and the consequent change in some social or business behaviours (e.g. e-working, e-commerce, e-conferencing, social networks, leisure time spent on the Internet) may affect mobility in different ways. It might lead to a reduction in travel demand as a number of activities no longer necessarily require physical travel or increase the number of trips and average trip distance, because relationships between people living farther away one from another are encouraged.

Thirdly, the knowledge of the compositional effect between drivers is partial.

Although some drivers have been investigated in detail it is usually impossible to define a universal mathematic law that correlates mobility and drivers.

The driving factors cannot be considered independently: a strong compositional effect should be taken into account and this effect may be very diverse considering different factors or different levels of the factors.

Also the relative importance of the factors may vary a lot.

To conclude, the recent changes in mobility have exposed the need to deal with a very uncertain future and have pointed out the necessity of updating the knowledge of mobility and revisiting the well-established planning models and the used approaches.

Further research should be carried out investigating changing transport trends and to evaluate the arising phenomena, such as the impact of new technologies and new patterns of economic and social development on accessibility and mobility. As a result, data collection—that could take advantage of the new opportunities given by technologies (such as social networks)—will assume a crucial role to provide better, more up-to-date information on mobility needs and patterns and should be revisited and conducted more frequently. Furthermore, the approach to planning should be adapted to respond better to the changing mobility demand and should be more integrated, flexible and responsive, for example responding to traffic congestion by implementing demand management programmes that can be adapted as needed if mobility demand changes.

## References

1. ACI–CENSIS (2012) XX Rapporto, Dove è finita l'auto?. Analisi di una crisi senza precedenti, Roma
2. AISCAT (2013a) Aiscat in cifre 2012. Roma
3. AISCAT (2013b) Aiscat informazioni, edizione mensili. Roma
4. ASFA (2013) Chiffres clés—Key figures 2013. Paris
5. Department for Transport UK (2011) Statistical Release—National travel survey 2010
6. Department for Transport UK (2012) Statistical Release—National travel survey 2011
7. Department for Transport UK (2013) Statistical Release—National travel survey 2012
8. EC Eurostat http://epp.eurostat.ec.europa.eu
9. EC (2013) Eu transport in figures—statistical pocketbook 2013. European Commission
10. EC (2008) Eu energy and transport in figures—statistical pocketbook 2007/2008. European Commission
11. Holl A (2007) Twenty years of accessibility improvements. The case of the Spanish motorway building programme. J Transp Geogr 15:286–297
12. Isfort (2011) La domanda di mobilità degli italiani. Rapporto congiunturale di fine anno
13. Litman T (2013) The Future isn't what it used to be. Victoria Transport Policy Institute, Victoria
14. Maggi S (2003) La cultura della mobilità in Italia. Storia e Futuro. Rivista di Storia e Storiografia 3
15. Sessa C, Enei R – ISIS (2009) EU Transport GHG: Routes to 2050? EU transport demand: Trends and drivers, project funded by the European Commission's Directorate-General Environment
16. TRANSvisions (2009) Report on Transport Scenarios with a 20 and 40 Year Horizon. Project funded by the European Commission—DG TREN, Final Report
17. Van der Waard J, Immers B, Jorritsma P (2012) New Drivers in mobility; what moves the Dutch in 2012 and beyond? Long-run trends in travel demand, OECD Roundtable
18. Weltevreden JWJ, Rotem-Mindali O (2009) Mobility effects of b2c and c2c e-commerce in the Netherlands: a quantitative assessment. J Transp Geogr 17:83–92

# Freight Transport: Main Issues

# 2

Gerard de Jong and Eric Kroes

Abstract

This chapter summarises some of the main issues in the analysis of freight transport. It starts by underlining why it is important to study freight transport, and which perspectives can be taken. Then the different types of agents involved are discussed, together with the choices that they have. These often lead to multi-leg transport chains, in particular when consolidation is involved. The chains can be analysed at microscopic or macroscopic level. Then the evolution of freight transport is discussed: what are the drivers of freight transport and what are the (expected) developments in those? The different modes of freight transport are compared, in particular road and rail transport, but also the other modes including water-based transport, air freight, pipelines and container based transport. Finally the policy levers available to influence freight transport are discussed.

G. de Jong (✉)
Significance and ITS University of Leeds, Koninginnegracht 23, 2514 AB, The Hague, The Netherlands
e-mail: dejong@significance.nl

E. Kroes
Significance and VU University Amsterdam, Koninginnegracht 23, 2514 AB, The Hague, The Netherlands
e-mail: kroes@significance.nl

I. M. Lami (ed.), *Analytical Decision-Making Methods for Evaluating Sustainable Transport in European Corridors*, Sxi 11, DOI: 10.1007/978-3-319-04786-7_2,

## 2.1 Introduction

There are all kinds of reasons for researchers (both theoretical and applied) to study freight transport. On the one hand this is a very interesting topic where the complex interplay of various agents shapes the transport of goods from the places of production to the places of consumption. On the other hand it is a key sector of a modern economy, that is vital for keeping the system going, but also one that plays a role in various policy debates (e.g. on the transport emissions, both local air pollution as well as greenhouse gases). The contribution of the logistics sector to GDP and employment in the EU is estimated to be around 7 %, which puts the logistics sector at a third place in the ranking of all sectors [1, 2]. Now freight transport is only a part of logistics. Beside the physical flow of the goods in space and time (freight transport), logistics is also about the service flow and information flow as well as about inventory management. In 2003, studies in the UK indicated that 4.8 % of the gross value added came from the service sectors' transport and storage and 7.8 % when also including communications [3]. Meersman and Van de Voorde [4] estimate that in 2009/2010 the share of the transport sector in the EU27 total value added was 4.4 % and in employment 4.5 %. However, in specific supply chains the transport and distribution of goods can account for as much as 25 % of the cost of products [5].

Freight transport can be studied from the perspective of a single firm, where the objective then is to improve freight transport efficiency so that the same amount of transport can be carried out at a lower costs, or a larger market area served (this is the viewpoint for many applications of operations research/business logistics). It can also be studied from a public policy perspective (the perspective of society as a whole): looking at freight transport patterns through the eyes of a national or regional government or an international organization of governments. This is the viewpoint we will take in this chapter. Another possibility that is becoming increasingly popular is to study freight transport at the level of a supply chain, involving several firms. The viewpoint of individual firms or supply chains will only be discussed in this chapter when this is relevant for our investigation of freight transport from the public policy perspective.

## 2.2 Agents and Choices Involved in Freight Transport

### 2.2.1 The Agents

One of the key differences between freight transport and passenger transport is that in passenger transport there often is only a single decision-maker for a journey, whereas in freight transport multiple agents are usually involved in the decision-making about a single shipment. The shippers (these can be producers or traders of commodities or their representatives) are firms that have a demand for transport services. In most cases these transport services refer to the activity of sending products to their clients, which are the receivers or consignees. In some cases, the receiver organises the transport. The shippers themselves, in what is referred to as own account transport, meet part of this demand. The remainder, hire and reward transport, is contracted out to carrier firms or intermediaries known as third and fourth party logistics service providers. Third party logistics (3PL) service providers perform logistics activities for a shipper, whereas fourth party logistics (4PL) service providers integrate capabilities of several organisations, including their own (e.g. multiple 3PLs for different parts of the logistics chain) to obtain a comprehensive supply chain solution.

### 2.2.2 The Choices

For freight transport, as for passenger transport, one can identify a set of choices which are made by relevant decision-makers that collectively determine the amount and composition of freight transport demand. These choices include [6]:

- Choices on production and consumption of goods and on trade and distribution. In most cases the underlying choice here is the sourcing decision; the decision of a producer, wholesaler or retailer from which supplier to buy the goods—this also determines the geographical location of the supplier and consequently the trade relation and transport needs.
- Shipment/inventory choices such as shipment size, frequency, etc. result in shipments of commodities with a certain weight, size, and value between the point-of-production and the point-of-consumption. The shipment's size and value are important characteristics because they affect the mode choice and the load factor. The load factor is the weight of the cargo divided by the capacity of the vehicle or vessel.
- Transport chain choices result in a series of modes and vehicle types used consecutively for a transport between the point-of-production and the point-of-consumption. This includes information on the transhipment(s) between the modes or vehicle types for the same mode. A chain contains a single leg using a single mode in the case of direct transport. It can also consist of several legs, each with its own mode or vehicle type, as depicted in Fig. 2.1. An example of a multi-modal, multi-leg transport chain would be: road transport from the point-of-production to a port, followed by sea transport to a second port, and finally road transport to the point-of-consumption. Transport chain choices include the choice on the number of legs in the chain, the mode choice for each leg and the transhipment location(s). These choices result in a modal split and affect the vehicle load factor. Together, the transport volumes, the mode shares and the load factor determine the number of vehicle-kilometres by mode.
- Finding return loads to avoid empty vehicle/vessel returns.
- Time-of-day choices and other timing issues such as the day of the week that produce a distribution of traffic over time periods.
- Route choices that yield the distribution of traffic over the network.

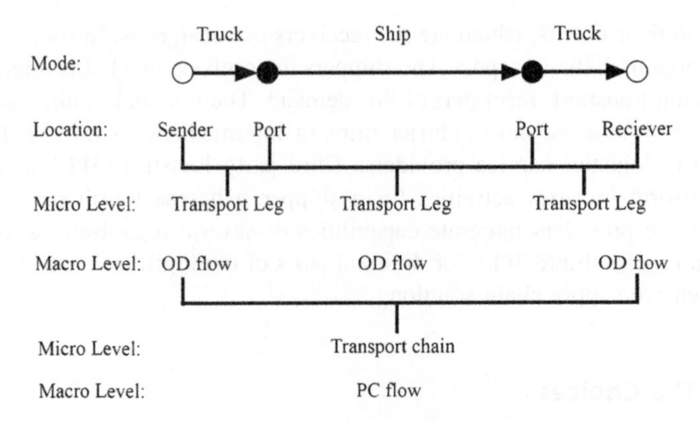


**Fig. 2.1** Transport chain, transport legs, PC flows and OD flows

In the context of international corridors, such as those distinguished by the European Commission, *all* these choices are relevant. Discussions about corridors sometimes exclusively focus on modal split, which is understandable given the potential that modal shift has for reducing external effects of freight transport. However, for understanding freight transport in such corridors, it is important to embed mode choice within the context of a transport chain all the way from producer to consumer and to consider sourcing, inventory an transport logistics (such as consolidation and distribution), timing and routing decisions as well.

Each of these choices can be analyzed on either *aggregate* data, which are the data at the macro-level of geographic zones, or *disaggregate* data, which are the data at the level of the decision-makers such as shippers and carriers. Within disaggregate data there is a distinction between data on actual choices in real life, referred to as revealed preferences, and data on choices in hypothetical situations, which are referred to as stated preferences. There are partly disaggregating national freight transport models in Italy, Germany and Scandinavia. But most freight transport models are aggregate models that are sometimes highly segmented by commodity type. The main reason for the lacks of disaggregate models in freight transport is the shortage in publicly available micro-data. This is related to the cost of collecting these data and the reluctance of private firms to disclose commercially sensitive information.

At the macro-level, a distinction can be made between PC (production-consumption) flows and OD (origin-destination) flows. This distinction is the macro-level equivalent of the distinction at the micro-level between a transport chain and a leg of a transport chain (also see Fig. 2.1). A transport chain can potentially be a multi-modal transport structure whereas a transport leg is a uni-modal structure, that is part of a transport chain.

By adding the volumes of the transport chains to and from the same zones, one obtains PC matrices. Similarly, by adding the volumes of the separate legs to and from the same zones, the OD matrices are obtained. PC matrices contain commodity flows all the way from the production zone to the consumption zone. These flows may consist of several OD flows, since a transport chain may be used with multiple modes and/or vehicle types as well as one or more transshipments along the route.

PC flows represent economic relations and transactions within different sectors of the economy and between these sectors. Changes in final demand, international and interregional trade patterns, and in the production structure of the economy, have a direct impact on the PC flow patterns. The available data on economic linkages and transactions are in terms of PC flows, not in terms of flows between producers and transshipment points, or between transshipment points and consumers. Changes in logistics processes such as changes in the number and the location of depots, and in logistics costs, have a direct impact on the composition of the transport chains. This can result in different OD flows, that would only indirectly impact the economic and trade patterns (and hence the PC flows). Assigning PC flows to the transport networks requires the transformation of these flows into their corresponding OD flows by mode. For instance, the transport chain road-sea-road would lead to road OD legs ending and starting at different ports. These would then be assigned instead of a long-haul road transport that would not involve any ports. A similar argument holds for a purely road-based chain which first uses vans to ship goods from the point-of-production to a consolidation centre, then consolidates the goods with other flows of goods into a large truck for the main haul, and finally again uses a van to deliver goods from a distribution centre to the point-of-consumption. In this scenario, the network assignment depends on the three OD legs. Therefore, converting the PC flows into OD flows, which can be done by a 'logistics model', [7] is required for a meaningful network assignment of freight transport. The data available for transport flows from traffic counts, roadside surveys and interviews with carriers are also at the OD level, not at the PC level.

The main reason for consolidation is the desire to reduce costs by sharing transport costs with other shipments. It allows the use of fuller and larger vehicles and vessels with lower unit cost. Consolidation can take place:

1. At the place-of-origin. A shipper can organise batches of products to be transported so that a good match with available transport capacity is obtained.
2. In consolidation centres that receive goods from several senders.
3. By means of collection vehicles that carry out collection rounds by visiting multiple senders.

Consolidation of type two or three requires multi-legged transport chains. This would include at least a leg from the sender to the consolidation centre as well as a consolidated leg from the consolidation centre to the consumption zone. Usually, the second leg will have a distribution centre as its destination. There usually is also a third leg from a distribution centre, where the consolidated shipment is broken down into multiple shipments for the different receivers. Consolidation can be done with storage at the transhipment location. However it can also involve the use of 'cross-docking', which means that the vehicle from which the goods are unloaded and the vehicle on which the goods are loaded are present at the terminal at the same time, and that the goods are moved from one vehicle to the other, without any storage. Cross-docking has increased substantially in recent years, and requires a high level of time-planning (enabled by modern tracking and tracing technologies).

### 2.2.3 Who Decides on What?

The relevant agents for production and consumption decisions are of course in first instance the sender and the receiver, but in the end the decisions are driven by consumer demand. Sourcing decisions are the domain of the receivers of the goods.

Managers of the shipper, the carrier and/or the intermediaries may make the transport choices such as the mode choice decision. Interactive decision-making (and especially research methods for dealing with this) is discussed in Holguín-Veras et al. [8] and de Jong [9]. In general, it is recognised that the shipper is the most common decision-maker for mode choice, also for transports that are actually contracted out to transport suppliers (carriers). Many carriers just offer a single mode, and in the case of a multimodal transport chain they may only be involved in a single leg of the transport chain (e.g. a road haulage firm that provides the first road transport in a road-sea-road transport chain). Logistics service providers on the other hand typically offer door-to-door transport services, and take over responsibility for the entire transport chain.

In most cases, the receivers are not responsible for organising the transport (deliveries to supermarkets can be an exception). But they usually are the key decision-maker concerning the moment in time that the delivery takes place, since they typically specify the delivery time window. This is also the case for the shipment size (and thus also the transport frequency) decisions, which are determined when they order the goods from the sender. The sender (and its transport suppliers) then has to take the delivery time and shipment size as given.

The firm that actually carries out the transport usually determines the route choice. In the case of road transport, truck drivers may have some freedom to choose the route or to change routes as a reaction to unexpected traffic delays.

A shipper or logistics firm can decide to make an integrated plan for a combination of the above choices; this is the topic of logistics network design, which is about finding well-balanced solutions in terms of the number of consolidation and distribution centres, their locations, the places where inventories are stored and the inventory levels and the transport organisation (modes, vehicle types, routes, departure times) [10, 11]. But is it also possible to distinguish between a group of long-term decisions (such as on the locations) and a group of short-term decisions (such as routing and departure time), that are conditional on the long-term choices. For several of the above choices, increasingly more powerful (commercial) software is available to support decision-making at the level of the firm (or supply chain), for specific operational problems (e.g. routing) all the way to more integrated and strategic decision-making (e.g. facility planning: number and location of warehouses).

## 2.3 Evolution of Freight Transport: Drivers of Freight Transport Choices and Developments in Those

Since World War II, in most years freight transport (measured in tons—kilometers) has grown at least as fast as gross domestic product (GDP). This is related to the fact that (international) trade has tended to grow faster than income and production and that the distances over which goods are traded have increased. Inland freight transport in tons—kilometers in the EU27 has followed closely the evolution of GDP in the period 1995–2007 [4].

Freight traffic (measured in vehicle kilometers) has grown even more than freight transport, partly due to changes in logistics systems as described below. To reach the societal objective of greater environmental sustainability, several governments and international organizations have placed decoupling the link between economic growth and freight traffic growth on the political agenda.

Before the most recent economic crisis, decoupling for road freight vehicle-kilometers was observed in some countries, such as Denmark, Sweden and the UK, but the opposite was still the case in other countries including Germany and The Netherlands [12].

After many years of growth, freight transport volumes (in tons—kilometers) fell in 2009 [13]. This goes for worldwide maritime transport (−6 %), worldwide air freight (−9 %), rail freight (especially in the EU27: −18 %), road freight (again especially in the EU27: −10 %) and inland waterways transport (EU27: −12 %). In 2010, there was a rebound for worldwide maritime and air transport, followed by a stabilization in 2011. Similar patterns for 2010 and 2011 can be observed for road and inland waterway transport in the EU27, whereas rail freight has grown both in 2010 and 2011 by 7 %. However, the US, Russia and China account for nearly 80 % of total estimated global rail freight transport [13]. In emerging economies, such as China and India, freight transport by road has been growing in all years in the period 2008–2011

In van de Riet et al. [14] and de Jong and Ben-Akiva [6] the key drivers of freight transport are identified. The most important drivers of total freight transport (measured in tonnes or km) are the volume and structure of consumer demand and production and the trade patterns. Logistic developments and attributes of the modes (especially costs, time, reliability, flexibility), on the other hand, are more important drivers of modal split and shipment size.

The following developments with respect to these drivers have taken place in recent years and can be expected to shape freight transport in the years to come:

- Consumer demand is likely to rise in many if not most parts of the world, which in turn would lead to an increase in the number of freight shipments. Furthermore, consumer demand is also likely to become more spatially dispersed (e.g. China, India, South America), which would lead to increases in transport distances.
- The above development will lead to an increase in trade among countries. But international trade will also grow due to globalisation of production. This is likely to lead to further increases in transport distances.

- Further dematerialisation in areas like mail, newspapers, and tickets, leading to a reduction in freight transport trips.
- Increase in e-shopping and home deliveries. This will lead to a transition from shopping trips to freight distribution tours (usually with small shipment sizes).
- Over the past few decades, logistics has changed dramatically due to greater competition in the logistics and transport markets that has been advanced by various technological innovations (mainly in ICT). Developments in the logistics systems being used, that have been going on for some time now, and can be expected to continue are:

  - Unit transport costs have decreased over the last decades while unit inventory costs have increased (only in recent years these trends have halted). The change in the relationship between storage and transport costs has been a major cause of the use of the just-in-time (JIT) concept, which has led to a decrease in inventory levels and shipment sizes and an increase in delivery frequency. This has resulted in an increase in vehicle kilometres and an increased demand for transport by van or small truck instead of heavy truck transport. The growth of JIT transport increases the service requirements of the transport modes, especially with regard to reliability of the transport time (delivery at the agreed time or within the agreed time window) and flexibility (short reaction time between order and delivery). The dominant perception among firms that require transport services is that road transport modes perform considerably better than other modes on these factors. To some degree the actual performance of rail, inland waterways and short-sea shipping might be better than perceived, but it is also a matter of natural disadvantages and possibly inefficiencies in the organisation of non-road transport. So, the growth of JIT transport has improved the competitive position of road transport.
  - Technological developments in production facilities and supply chains are facilitating demand-driven production. Two components can be distinguished here. The first component is lean production—this is the flexible production of (semi-)manufactured goods, whereby the production facility can be reconfigured within hours (instead of days) to switch between products. This enables manufacturers to produce a wide range of products and a wide diversity of a given product at a single facility. The second component is postponement manufacturing. Semi-manufactured goods are produced according to a demand forecast (BTS: built to store) at a central production facility and are shipped to assembly facilities near the market. At the moment a final product is ordered, it can be assembled at the assembly facility, resulting in very short lead times and quick fulfilment. Due to the production of a variety of components, orders can be customised to match the demands of the customer. This influences what needs to be shipped and where it is shipped, and these developments put specific demands on the supply chain. The supply chain must be flexible enough to enable short lead times (time between order and delivery) and also to enable a reliable delivery of products. ICT tracing and planning systems facilitate the control of material flows,

providing real-time information on the status of the products. This has resulted in a restructuring in the management of the supply chain. The various transport modes differ in the way they can meet the demands for shorter lead times and JIT delivery. Shippers usually view road transport as the mode that can provide the highest flexibility and reliability.

- Another important development in supply chain management is the increased use of distribution centres and of hub-and-spoke systems. This helps firms to reduce the costs of distribution facilities, transport, warehousing, and inventory. Economies of scale can also be achieved by concentrating production facilities in fewer locations and by centralising inventory through a reduction of the number of stockholding points. Inventory centralisation nowadays occurs on a larger geographical scale than before, which results in longer routes in general, but also in a consolidation of traffic flows. Consolidating freight flows leads to higher load factors, use of larger vehicles, and opportunities for alternative modes (rail, inland waterways, short-sea shipping) on the long haul. Larger vehicles are more economical in terms of cost per tonne than smaller ones, provided they are fully loaded. By consolidating freight flows, it is possible to collect sufficiently large volumes for transport over longer distances by vehicles of a larger size. Furthermore, consolidating freight flows, especially in combination with a trend towards more containerisation and an increase in global trade volumes, makes non-road transport a more attractive option for the long-haul. Comparing this development to the two mentioned directly above, implies that we expect changes working in opposite directions; however the spatial scale is different: bigger vehicles are used more and more in hub connections for long distance (national, international) transport, whereas smaller vehicles are used more and more in urban and regional distribution. This favours road transport, unless road congestion problems increase.
- Shared use of transport and warehouse facilities that will lead to higher vehicle load factors.
- A further use of logistics planning systems, tracking and tracing and real-time information. This will lead to higher load factors and use of less congested routes and time periods.

Some of these developments are increasing the future freight volumes and some are decreasing them. Nevertheless, based on recent trends, it is much more likely that the former developments will dominate and that freight transport and traffic will continue to increase during the coming decades.

## 2.4 A Comparison of the Modes of Freight Transport

In terms of freight transport modes, the available options generally are road, rail, inland waterways, sea, air and pipeline. Within these modes, several types of vehicles or vessels, such as articulated trucks or solo trucks, can be distinguished. Road transport is generally the most widely available mode. The availability of inland waterways modes and short sea shipping is the most constrained. The characteristics of the different modes are discussed below.

Rail Networks have a Much Lower Density than Road Networks and Only a Few Firms have Direct Access.

Railway operations often require reconfiguration of trains at marshalling yards, which is time-consuming and leads to relatively long door-to-door transport times. In this regard, three different rail products can be distinguished:

- Full train loads between two private sidings requiring no remarshaling or transhipment. Such services require very large consignment sizes, but then will have low costs at any length of haul.
- Wagonload services which require remarshaling to consolidate traffic; these are only viable for large flows over long distances.
- Container or other intermodal services which generally use consolidation and access by road; these require long distances (although traffic to or from ports can serve a terminal at the port, and do not require transfer by road, except perhaps within the port. Thus these are more favourable for rail than other intermodal traffic).

Rail transport requires very substantial investments in tracks, signalling, terminal facilities and equipment, some of which can be shared with passenger transport (but then the capacity also has to be shared with passenger trains, which often get priority, especially in Europe). Because of these substantial fixed costs, the unit cost of transport is high at low transport volumes and decreases slowly with increasing transport volumes. Rail transport is often the least cost choice for large quantities of goods transported over long distances. For this reason, providing rail transport was historically regarded as a natural monopoly and even today a single entity usually manages the rail tracks of a country, although operations are separated from the management of the tracks. However, in many countries rail deregulation and privatisation have taken place. This has resulted in competition between freight transport rail operators.

In road transport, labour expenses are the main component of transport cost. The fixed costs are a considerably less significant portion of the total cost than that of rail transport, partly because it uses all-purpose roads supplied by the state. In most countries, the road haulage sector is also highly competitive, consisting of many firms varying in size from large corporations to one-person owner-operators.

In Table 2.1 the pros and cons of the rail and road systems are compared. This table is based on current practices in Europe instead of theoretical characteristics of the modes. In some countries, such as the U.S. where freight transport by rail usually gets priority over passenger transport by rail, the pros and cons of the two systems may be somewhat different. Road transport usually scores best on: time, reliability, flexibility and accessibility. Conventional rail and combined road-rail transport (intermodal transport) have relatively better safety and cost features especially for long distances and/or large volumes. In cases where there is a high load factor, inter-modal transport also produces lower emissions of conventional

**Table 2.1** Strengths and weakness of road versus rail in EU freight transport

| Mode | Strengths | Weaknesses |
|---|---|---|
| Railway | Adequate service level for bulk | Less innovative (information systems) |
| | Direct transport between large-volume centres | Compatibility in international transport |
| | Safety | Time and cost for loading and unloading and marshalling (if needed); limited opening hours of facilities |
| | Low emissions | |
| | Price (long distance, large volume) | Bottlenecks on some links due to competition with passenger trains |
| Truck | Speed | Higher emissions |
| | Flexibility, timely available | Capacity bottlenecks, congestion risks (also due to competition for road space with cars) |
| | Spatial coverage | |
| | Possibilities for consolidation-en-route | Increases road maintenance costs |
| | Small consignments | |
| | Point-to-point shipments | |
| | Quality of handling | |
| | Information systems | |
| | Transport time reliability | |

pollutants and greenhouse gases. Rail freight transport has the potential to be very reliable. But in reality, especially in Europe, shippers usually consider the timeliness of road transport to be superior; it is difficult for rail to hit tight delivery windows. If road congestion would further increase, this might change in the future.

Train operations are also potentially less sensitive to weather conditions than road transport. International rail transport in Europe is still slow and costly due to the lack of interoperability and responsiveness to market forces dictated by national railroads. It can only remain competitive in long distance transport routes over 350–500 km [15]. Therefore, rail transport is commonly used to transport low value bulk cargo where the most important factor is low rates.

In order to obtain substantial rail market shares in other cargo, a truly intermodal system with one logistics service provider that is responsible for the entire transport chain while offering reasonably fast and reliable door-to-door services would be required.

Waterways and sea routes generally require no infrastructure investments from the carriers since public authorities usually undertake these costs. Carriers only invest in the equipment [16]. The sea transport market is characterised by several large private carriers competing for international market share while smaller carriers compete for inland waterways transport.

Water-based transport vessels tend to be considerably bigger than the vehicles used in road and rail transport. They offer low unit rates but also low operational speeds. This makes water-based transport most suitable for the transport of low value goods over long-distances. Inland waterways transport competes with road, rail and short-sea shipping. Ocean shipping has a large market share when one measures this by shipping volume in tonnes. However measured in terms of value, a high percentage of international transport goes by air.

Passenger transport planes are still used for freight transport in what is known as belly freight, whereas airfreight transport via dedicated freight carriers is increasing (e.g. for express freight). Air transport has high variable costs compared to the fixed costs [16]. In comparison to other modes, the unit costs are quite high, and especially so for large volumes. For long distances, air transport is substantially faster than all other modes and also considered relatively reliable. Given its high costs, air transport is predominantly used for goods that quickly deteriorate in value or that are urgently needed at the destination. Some examples are flowers, newspapers, spare parts, and express mail. But, more recently, the air transport market has been expanding to other goods.

Pipelines are only used for liquid petroleum and natural gas. The investment costs for a pipeline network are very substantial and cannot be recovered when only transporting small volumes. These costs decrease as the quantity increases. Pipeline networks have low density and low speed.

Intermodal and multimodal transport both use several modes in the same transport chain between point-of-production and point-of-consumption. The main difference is that intermodal or 'combined' transport is carried out for a single flat rate and uses the same loading unit and volume on all the modes in the chain as opposed to multimodal transport. This unity reduces the transhipment costs and time as long as specialised equipment for transporting the containers at intermodal stops are available. The most common containers are eight feet wide and twenty feet or forty feet long. Container movements are often measured in TEUs: twenty-foot equivalent units. The use of containers began in the maritime sector initiated by the Sea-Land Company, and it has grown tremendously over the past decades. Containers can be used on sea vessels, trucks, single and double stack trains, as well as inland vessels. Road-sea-road and road-rail-road are transport chains that are used regularly using containers all the way. Intermodal transport can also use swap bodies that have non-rigid sides, instead of containers. Besides container ships, there are also roll on–roll off ships, called RoRos, where the road trailers are driven. Trailers can also be loaded onto trains (this is sometimes referred to as 'Rollende Landstrasse', 'Iron Highway').

Overviews of demand elasticities for the impacts of price changes in roads and rail transport can be found in de Jong et al. [17] and VTI and Significance [18].

## 2.5 Policy Measures

The public authorities' main objectives for transport and land use policies with respect to freight transport demand are to promote accessibility and decrease the negative external effects of transport. Apart from this, safety and security in transport can also be important reasons for government intervention. Among the key drivers of freight transport demand, the most important drivers are growth in the gross domestic product (GDP), growth in trade and the increasing physical distance between suppliers and receivers. International trade has been growing considerably faster than global GDP. In Europe, the creation of the single market has contributed to the growth of freight transport. Logistic developments such as the centralisation of inventories, emergence of hub ports and airports, reduction of lead time (as part of just-in-time supply chains) and increase of business-to business electronic information exchange [19] and the performance of the modes (in terms of the attributes costs, time, reliability and flexibility) are important drivers influencing modal split and shipment size.

It is clear that the main drivers, such as growth in GDP and growth in trade, are outside the span of control of transport and spatial planning policies, which typically are about changes in transport time and cost by mode and restricting certain developments in specific zones. Hence such policies can only have a modest impact on the volume of freight transport. But transport and land use policies can still have a significant impact on the volume and composition of freight transport in the following way [6, 14]:

- These policies can affect residential and business location choices. For example, regulations or financial incentives to businesses can stimulate mixed land use or spatial clustering, which has the potential to reduce transport distances.
- Governments can promote certain types of logistic developments. For example, they can encourage the emergence of information brokers and other means of providing data on consignments, in order to increase the logistics efficiency by increasing load factors and decreasing empty backloads.
- Transport policy can influence some of the characteristics of the modes, such as transport time, by investing in network links or in public or public/private terminals (e.g. as part of city logistics). They can also affect cost by implementing taxes, tolls and subsidies. Note however that several logistics costs components have already been increasing considerably in recent years, due to fuel price increases, driver shortages and as a response to environmental concerns. In some segments of the freight transport market such as low value goods and long distance shipments, the modal split is rather responsive to time and cost changes by mode. Cost increases might also lead to increases in the efficiency of transport (higher load factors, less empty driving) and be an incentive to choose more nearby suppliers. The scheduling of freight transport and of distribution in particular may be influenced by delivery time windows, time-of-day specific vehicle bans and by time-of-day specific road pricing. These

restrictions are more effective when applied in combination with incentives for the receivers to change delivery times to periods outside the normal business hours. The choice of vehicle type can be influenced by vehicle weight and height restrictions, emission standards or load factor requirements in specific areas such as city centres.

- Introducing more competition into the rail freight sector through government policies might lead to improved service characteristics such as a more innovative organisation, improved reliability, interoperability, and flexibility of the rail freight sector vis à vis the road sector.

In summary, the most effective policies for affecting freight transport demand would appear to be policies focused on changing transport mode characteristics and spatial planning policies. Such policies have the potential for producing significant changes in the (composition of) freight transport demand. In particular, they can shorten transport distances and improve the way the transport system is used (by affecting the choice of mode, time-of-day, and route). Some of these policies will take a long time to implement and for their results to be realized (such as land use planning and infrastructure expansion), while others can be implemented more quickly and will produce their desired effects in the short term (such as road pricing). Because large shifts in demand will be needed to produce a sustainable freight transport system and time is of the essence, both short-term and long-term actions should be considered.

## References


1. PWC, IWW (2007) Preparatory study for an impact assessment on a EU freight logistics action plan. Study for the European Commission, PWC, Rome
2. Rothengatter W (2010) Trends of logistics after the crisis. In: van de Voorde E, Vanelslander T (eds) Applied transport economics. A management and policy perspective, De Boeck nv
3. National Statistics (2007) United Kingdom national accounts. The Blue Book, National Statistics, London
4. Meersman H, van de Voorde E (2013) The relationship between economic activity and freight transport. In: Ben-Akiva ME, Meersman H, van de Voorde E (eds) Freight transport modelling. Bingley, Emerald
5. Heizer J, Render B (2007) Operations management, 8th edn. Person Prentice Hall, New Jersey
6. de Jong GC, Ben-Akiva ME (2010) Transportation and logistics in supply chains. In: Bidgoli H (ed) The handbook of technology management. Wiley, New York
7. de Jong GC, Ben-Akiva ME (2007) A micro-simulation model of shipment size and transport chain choice. Spec issue freight transp Transp Res B 41:950–965
8. Holguín-Veras J, Xu N, de Jong GC, Maurer H (2011) An experimental economics investigation of shipper-carrier interactions on the choice of mode and shipment size in freight transport. Netw Spat Econ 11:509–532
9. Jong GC de (2012) Application of experimental economics in transport and logistics. European Transport/Trasporti Europei 3(50)
10. Chopra S (2003) Designing the distribution network in a supply chain. Transp Res 39:123–140
11. Chopra S, Meindl P (2006) Supply chain management: strategy, planning and operation. Pearson Prentice Hall, New Jersey

12. Kveiborg O, Fosgerau M (2007) Decomposing the decoupling of Danish road freight traffic growth and economic growth. Transp Policy 14(1):39–48
13. International transport forum (2013) statistics brief Dec 2013, Trends in the transport sector: global transport trends in perspective, ITF/OECD, Paris
14. van de Riet OATM, de Jong GC, Walker W (2008) Drivers of freight transport demand and their policy implications. In: Perrels A, Lee-Gosselin M (eds) Building blocks for sustainable development. Bingley, Emerald
15. Beuthe M, Kreutzberger E (2001) Consolidation and trans-shipment. In: Brewer AM, Button KJ, Hensher DA (eds) Handbook of logistics and supply-chain management. Handbooks in transport, vol 2. Pergamon, Amsterdam
16. Coyle JJ, Bardi EJ, Langley CJ (1996) The management of business logistics. West Publishing Company, St. Paul
17. de Jong GC, Schroten A, van Essen H, Otten M, Bucci P (2010) The price sensitivity of road freight transport—a review of elasticities. In: van de Voorde E, Vanelslander T (eds) Applied transport economics, a management and policy perspective. De Boeck, Antwerp
18. VTI, Significance (2011) Priselasticiteter som underlag för konsekvensanalyser av förändrade banavgifter för godstransporter. Report for Banverket, VTI/Significance, Stockholm
19. Ruijgrok C (2001) European transport: insights and challenges. In: Brewer AM, Button KJ, Hensher DA (eds) Handbook of logistics and supply-chain management. Handbooks in transport, vol 2. Pergamon, Amsterdam

# The Evolution of Maritime Container Transportation

**3**

Bruno Musso

Abstract

The significance of maritime container transport is well known today, as is the importance it has had in the birth of the global village and how the drive for internationalisation has produced the immense traffic flows that feed it. The container is one of the many standardisation processes that make the industrialisation of maritime transport of general cargo possible. This form of transportation is used worldwide and therefore requires standardisation in all terminals in the world and for all means of transport. A great effort is being made to unify international standards, which are controlled by special organisations. But the need for standardisation inevitably clashes with the need for change dictated by technological evolution. The conflict between evolution and standardisation, somewhat common to all industries, is particularly accentuated in the field of global container transport and is conditioning its nonetheless explosive development.

## 3.1 Introduction

The significance of maritime container transport is well known today, as is the importance it has had in the birth of the global village and how the drive for internationalisation has produced the immense traffic flows that feed it. The container is one of the many standardisation processes that make the industrialisation

B. Musso (✉)
Grendi Trasporti Marittimi Spa, Genova, Italy
e-mail: bruno.musso@grendi.it

I. M. Lami (ed.), *Analytical Decision-Making Methods for Evaluating Sustainable Transport in European Corridors*, Sxi 11, DOI: 10.1007/978-3-319-04786-7_3,

of maritime transport of general cargo possible. The process originates from the standardisation of a trailer box that is made separate from its wheels and is fitted with specific features to facilitate handling by mechanical means. These standardised coupling systems, the corner fittings, not only enable automated handling, but are also used for stacking up to 9 containers and for lifting them onto the means of transport, be it ship, train or trailer. As a result, "general cargo" becomes one single, homogeneous object expressly designed to be moved from one mode of transportation to another, with similar procedures, costs and time frames as those for handling bulk solids (iron ore and coal) and bulk liquids (petroleum).

The idea was conceived in the US in the 1960s, by the owner of a transport company, Sea-Land, whose trucks connected the east coast with the west coast of the US. He thought of separating the trailer-box from its wheels in order to embark it on the ships. However, the innovation struggled to establish itself, because firstly this was not about an industrial standardisation process of one's own production process inside one's own company, but about transportation that involved various areas of the territory, often public ones (ports, roads), and secondly because it was in conflict with many interests that would have been compromised by the new technology. A perfect example was the opposition of the dock workers, whose activities were almost completely eliminated by the new technology and who for a long time tried to prevent its introduction with strikes and other forms of protest; in Italy this continued into the 1990s.

The new technology did, however, establish itself during the war in Vietnam, when container transport offered the only means to adequately supply the US army; this set the beginning of an epochal transformation that would revolutionise maritime transportation and reshape the global economic geography.

But that was not the end of the difficulties encountered by the new technology; the problems had only just begun and still affect current evolution today. In fact, this form of transportation is used worldwide and therefore requires standardisation in all terminals in the world and for all means of transport. A great effort is being made to unify international standards, which are controlled by special organisations. But the need for standardisation inevitably clashes with the need for change dictated by technological evolution. The conflict between evolution and standardisation, somewhat common to all industries, is particularly accentuated in the field of global container transport and is conditioning its nonetheless explosive development.

## 3.2 Elements of Constraint and Evolution in Maritime Transportation

### 3.2.1 The Container

When analysing the main aspects in the evolution of the container, we should start with the coupling and handling mechanisms: the corner fittings have remained incredibly unvaried for over half a century, proving that the initial solution was extremely efficient and did not need any further updates.

It is a different story all together with container sizes that have continued to vary according to the transportation mode involved and represent one of the big constraints. Let's look at the different sizes:

- *Height*: has continuously increased from initially 8 ft (244 cm) to 8.5 and 9 ft and up to 9.5 ft; however, this change did not cause major problems because the ships, except for some special cases, adapted to the new sizes and, if necessary, loaded one less row of containers, without great repercussions on structures and organisation.
- *Length*: obviously had to adapt to the increasing trailer size; the first containers measured 27 ft (Matson) and 35 ft (Sea–Land) and then reached a standard length of 40 ft (12.20 m—with submultiples of 20 ft). This length remained unvaried for many years and has become the dominant measure at present. But the evolution of trailers didn't stop; in the 1970s, the 40-ft trailer exceeded the maximum length permitted in Italy, whereas today the larger size of 45 ft, or 13.60 m, is allowed in the EU and US, but shipping restrictions stand in the way. In fact, a change in container length makes it necessary to also change the cell guides and the deck fittings of ships and to find solutions to accommodate different container sizes that cannot be calculated in advance. For this reason, there is a tendency to build 100-ft cells that are compatible with 20-, 30-, 40- and 45-ft containers, with little loss of cargo space. This, however, is possible only in new ships and still, some compatibility problems remain.
- *Width:* this is the big constraint that has not yet been solved and that, though little known of, limits the use of the container considerably. The container has Anglo-Saxon origins and is measured in feet, whereas continental Europe uses the metric system, and in this case the two measures are not compatible. The width of a container is 8 ft, that is 2.44 m, offering 2.36 m free inner space; the maximum allowable width of trailers in Europe is 2.5 m, with a useable inner space of 2.44 m. As long as cargo loading was carried out manually, the difference of 6 cm was marginal, but in the 1970s the handling of goods in warehouses and means of transportation ceased to be manual and instead was performed by mechanical means, such as forklifts, and hence the goods had to be positioned on pallets. European pallets are of rather standard dimensions, measuring multiples of 10 cm (also the packaging sizes), the prevailing standard being 120 × 80 cm; the others measure 120 × 120 cm or 80 × 80 cm or also 100 × 120 cm, but this ultimately does not change the problem.

In fact, all these sizes are coherent with the trailer inner space of 2.44 m, because it can fit 2 or 3 pallets side by side, but they are not so with the container, where they do not fit because of the 6 cm difference. In order to use the container, it is therefore necessary to either load the goods manually, taking the pallet apart, or to under-utilize the container with irrational solutions (120 + 80). This is not only a waste of space, but also requires complex solutions to fill the empty spaces in order to prevent the goods from moving inside the container during shipment thus causing enormous damage. Trade fairs are full of such complicated and expensive contraptions. As a result, the container is not useable for traffic within Europe, e.g. Milan–Paris, because there is another option, the trailer; so the transport operator drops the more logical solution, the container train, in favour of road transportation which is more irrational, but saves him from having to deal with the problem of loading the container. In spite of unrealistic proclamations, this restricts the development of the container and limits potential savings from increased rail transportation, while it augments pollution and saturates Europe's road network. With the Directive of April 7th, 2003—COM (2003) 155 final—Directive of the European Parliament and of the Council on Intermodal Loading Units—Europe tried, without success, to impose a container that is europallet compatible; but changing the container width requires ships to adjust their cell guides and the lashing points on deck. In the new ship constructions of today some changes have been made, but the problem of normal containers in the wider cells remains partly unsolved.

### 3.2.2 The Ship

We know that the optimal size of a ship increases with increasing cargo-loading speed and voyage length. In long-distance ocean transport, the increased cargo loading/unloading speed makes it convenient to use ships of ever larger dimensions; this phenomenon has long been observed in the transportation of bulk solids and bulk liquids, where the size of ships has grown continuously, reaching dimensions of around 300,000 tons. The use of containers has increased the speed of loading/unloading and of transporting general cargo more than 10-fold, arriving at the same levels as for bulk goods. A good berth for large ships can handle over 6,000 TEUs a day (more than 70,000 tons).

This has lead to the use of increasingly larger ships, passing from the 600 TEU ships of the 1970s to the 18,000 TEU liners that are starting to enter service; this phenomenon is driven by the strong economies of scale of a ship, with the costs of construction and crew increasing far less than the transport capacity. However, there is yet another particular multiplier of hydraulic nature, because at sea only little power is required to move a ship (big sail ships used to be towed by rowing boats when there was a lack of wind); the need of power then grows proportionally to the speed desired.

This proportionality continues until the optimum speed has been reached for any type of vessel, after which any further increase in speed requires a more than proportional increase in power up to its limit point, beyond which it becomes almost impossible to increase any further. The optimum speed depends on a whole lot of elements linked to the vessel, but one of the key factors is its length. The increase in ship dimensions is therefore not only linked to the economies of scale, but also—and especially—due to the tangible energy savings, which, as we will see later on talking about competition, represent one of the crucial elements in the competitiveness of shipping companies.

These large ships need to be supplied with immense volumes of traffic and this has been made possible thanks to the "global village", which was cause and consequence of this worldwide organisational evolution. In the past 40 years, European container traffic has increased more than 70-fold, much more than the GNPs [4], concentrating in the big ports of Northern Europe, which is a result of and also the reason for the strong concentration of production in that area. Today these ports handle cargo volumes of 10 million TEUs. Of course not all ports have a hinterland capable of generating such traffic volumes and therefore marine transshipment centres have been created along the routes of the giant ships. In these hubs the cargo of the large, so-called "mother" ships is transferred by feeder services onto smaller, 2.000–3.000 TEU ships ("daughters") that serve the other ports. Hub ports have less traffic, but are strategically positioned in respect to the markets they serve: the main hubs in the Mediterranean are Gioia Tauro, Malta, Algeciras and Tangier and the feeders serve all the Mediterranean coast of both North Africa and Southern Europe, since the Mediterranean regions surrounding the various ports are rather small and generate freight traffic of only about a million TEU. The Po Valley in Northern Italy is the only exception, as it has a market comprised within a radius of 150 km that produces an import/export volume of about 7 million TEUs; however, the maritime traffic serving this flow of goods is not concentrated in one big, strategically located port, but is split between 8 ports in Upper Italy, 4 of which in the Tyrrhenian Sea (4 million TEUs) and 4 in the Adriatic (1 million TEUs), and the traffic arriving from the ports in Northern Europe. This inadequate port situation, the very antithesis of the evolution needed, does not allow the use of mother ships and imposes the use of feeder service with daughter ships, thereby causing the negative effects we will see when we talk about territory. Only recently has Italy taken note of the large ships, realising that they cannot enter the Italian ports due to their dimensions and draught. They are trying to remedy the situation now, especially in Genoa, with projects to transform the ports making them suitable for new ship dimensions. However, ship size is only one aspect of the new requirements asked of the supply chain, because berth mustn't only offer good approach operability (dimensions, draught), but it also requires operational adaptations; a berth must be able to move 1.5–2 million TEUs a year, almost the same as the entire traffic in the port of Genoa. This requires an area of 150–200 hectares (all of the port of Genoa) and land transportation structures (road and rail) that at present do not exist and are unlikely to be built; this means a lack of outflow capacity to serve this hypothetic new traffic in addition to the existing one. The contemplated projects can therefore only have a nautical value; they lack consistency in logistics and are destined to trigger useless debates about plans that are not feasible.

Returning to ship dimension, we can observe that the medium size of ships has increased significantly over the past ten years, arriving at 13,000 TEUs, and this trend is continuing in the ships under construction or in delivery, with a predominance of ships of 18,000 TEUs. The specialists, however, tend to further extrapolate the trend towards increasingly larger ships, without considering that from a certain size on it is likely that scale diseconomies will prevail. The same mistake was made about 20 years ago, when the phenomenon of size increase was first seen with oil tankers. Also back then one had speculated with a continuous growth, to the point of designing an oil tanker of 1 million tons. Understanding the absurdity of such a dimension, the project was set aside and the maximum dimension remained around 300,000 tons. Not much later, the phenomenon of increasing dimensions also involved ships carrying solid bulk cargo (iron ore and coal). Also in this case a maximum size limit was determined, around 400,000 tons, which is slightly higher than that for oil tankers; probably because of the lower value of the minerals transported and the lower risk of marine pollution.

It is believed that a similar limitation should also apply to container ships, with a maximum dimension of around 18,000 TEUs, which is a little lower than for ships carrying bulk cargo. This is reasonable considering that goods transported in containers are semi-finished and finished products, therefore of high value, and require urgent delivery. So size limitations are not only dictated by the progressive economies of scale, but also by transportation requirements that advise against concentrating the cargo load beyond certain limits. Contrary to what is often said, we are therefore convinced that the ships capable of serving a final destination port in the Mediterranean will not exceed 18,000 TEUs.

In confirmation of this it should be remembered that after having increased the size of their ships, the biggest shipping companies, and hence the ones most susceptible to the evolutionary needs, like Maersk, opted to increase the frequency of departures in order to enhance their competitiveness, establishing a daily service between the big North European ports and the Far East. It is easy to understand that a daily service reduces transit times and port costs considerably and renders the entire logistics cycle smoother and more regular, thereby generating high savings that outweigh the economies of scale of ships by far: in fact, with the current ship sizes, the economies of scale have little effect and are counter-balanced by the diseconomies of scale of navigation (dimension and draught) and the whole logistics cycle.

### 3.2.3 Territory

The afore-described evolution has a strong impact on the territory, which needs to be organised adequately to prevent the territory from being overwhelmed by it. The first impact regards production possibility, where a hierarchy is generated between regions with final destination ports that are served by the mother ships and the others. Smaller ports moving 1 or 2 million TEUs are served by feeder ships arriving from the transhipment hub ports (in the Mediterranean Gioia Tauro, Malta, Algeciras, Tangier etc.) and this entails higher freight charges and transit times. In the global village, where competitiveness is a matter of a few percentage point difference in cost, such a handicap can kill a region's competitiveness. Especially when dealing with a modern style of organisation which is based on the just-in-time principle and characterised by strong financial constraints, a region can lose its competitive edge if it suffer delays (even just a week) in supply and sales.

A further constraint regards the cargo carrying capacity of the different modes of transport: while ship sizes have grown 70-fold in the past 40 years, from 600 TEUs to 18,000 TEUs, trailer characteristics have remained unvaried, except for some minor changes in shape, and also the size of trains has not increased significantly, particularly in Europe. So while 40 years ago one ship equalled 300 trailers and 12 trains, today it corresponds to 9,000 trailers and 360 trains.[1] In other words, the great development in cargo traffic that has taken place over the past 40 years is only compatible with the capacities of maritime transportation, but not with those offered by land transport, be it road or rail. The following figures support this assertion: on a regular day including the peaks, a 10 million TEU port, like those in Northern Europe, produces a daily traffic volume of 50,000 TEUs, which alternatively corresponds to 25,000 trailers or 1,000 trains (Genoa moves 25).

In terms of cargo-carrying potential, these quantities are not compatible with land transport. As for fluvial transportation (possible only in Northern Europe), however, transport capacities have increased considerably, while pollution has decreased at the same time. Fluvial transportation nevertheless remains an additional burden in terms of cost as well as of transit time from port to destination; it is a situation comparable to when a final destination is reached via a feeder service. The best solution is territorial rationalisation, with the production facilities located close to the port in order to minimize transfers in the region.

As a matter of fact, this happens almost automatically, because in advanced systems such as those crossed by the Genoa—Rotterdam corridor, production basically consists of receiving semi-finished goods, assembling and processing them to then send them back to the market. Delivery and dispatch are largely carried out by sea, so to be located in the close vicinity of the port brings significant savings in cost and time. The port therefore acts as a magnet for rationalising territory and reducing costs and land transport.

The ports of Northern Europe are a perfect example for this: their large surface area of around 10,000 ha is used only to a small extent for port handling; the main part of the area is used for production, storage, deposits, manufacturing, commercialisation etc. More than 50 % of the goods arriving in Rotterdam and Hamburg find their final destination within a range of 50 km from the port. The remaining, already considerably reduced amount of goods is then shipped on efficiently, not only by road, but depending on their destination also by rail and inland waterways. Land transportation persists thanks to this expert combination of factors, the only one that makes it possible to handle the final destination traffic of mother ships and thus establish a direct connection to the entire world.

In the Mediterranean the situation is a different one: almost all the various economic districts are of limited size and characterised by a hinterland that produces traffic of not much more than a million TEU and therefore can only be served by secondary feeder vessels. In this scenario land transportation is limited, since there's the tendency to establish oneself in the proximity of the only major port, but freight costs are higher because of the reduced quantities and the double

---

[1] To give the reader a clearer visual, consider that 9,000 trailers lined up on the motorway cover about 900 km, which is about 80 % of the distance between Genoa and Rotterdam.

journeys of the feeder ships. This is the inevitable price the periphery regions have to pay to the economies of scale of the global village.

The only exception is Northern Italy, where about 7 million TEUs are consumed or produced in a single hinterland comprised within a range of 150 km. This volume is the only one compatible with the use of mother ships; however, it is not concentrated in one single port, but split between 8 ports in Northern Italy and 2–3 ports in Northern Europe, because it has not been possible to build a port with the necessary characteristics. In comparison to the 10,000 ha-port complexes in Northern Europe, Genoa, which is the largest Italian port and only possible candidate for such function, covers only 250 ha and has the Apennine mountain range behind it that is difficult to cross. Not even the much discussed "Terzo Valico dei Giovi" project will be able to solve this problem, because of the excessive time it will take for its completion and because its capacities are absolutely inadequate; the railway operators themselves have not projected to carry more than a million TEUs, a volume that objectively speaking is rather unlikely and might have been substantial 40 years ago, but is definitely insufficient if we are looking at a future with ships of 18,000 TEUs and cargo traffic of around 10 million TEUs.[2]

The widespread hope/fear that the opening of the Alpine crossing could make it easier to supply the Po Valley from the ports in Northern Europe is also unfounded, because a rail transport that crosses all of Europe from north to south lacks potential to be able to cope with the expected increase in traffic volume and is presently already submitted to ecological constraints and showing signs of saturation.

There is little concrete data on the present traffic destined to the Po Valley and coming from the ports in Northern Europe and various studies effected show rather contrasting results; most estimates, however, value it to be a little less than 2 million TEUs a year. This means more than 150 trains a day, based on the average Italian train size; probably they are a few less, because European trains have more freight wagons, but that does not change much regarding noise pollution and saturation of the railway line. It is therefore hard to believe that Europe will be willing to accept a further increase in transit traffic caused by the inefficiency of Italian ports. Already now, some countries like Switzerland and Austria have introduced restrictions in order to limit transfer traffic through their territories. So far these restrictions only concern road traffic, but if the present traffic

[2] An entirely different and explosively innovative project for a final destination port for future large oceanic ships is the one designed by SiTI of Turin (Politecnico and Compagnia di S. Paolo) together with a group of operators from Genoa and called "Bruco"(Bi-level Rail Underpass for Container Operations). The project foresees a capacity of up to 10 million TEUs and occupies an area of around 500–800 ha in Piedmont, where other 3,000 ha of intermodal centres already exist in the neighbourhood. This area can be connected to the docks of Genoa Voltri through a type of conveyor belt with its own tunnel; the water depths alongside the quays will be 20 m and in using also the existing break water, the port will have a quay length of 3,200 m to accommodate mother ships (7–8 berths) and additionally 4 berths for feeders, ferries and container ships in the Mediterranean. The project could be auto-financed, but is still in the planning stage (for more detailed information see [1, 2, 4].

growth rate continues, it is likely that they will also involve rail traffic; and they will definitely first apply to traffic that is not related to local production needs, but is generated by an operational dysfunction in a neighbouring country.

This obviously results in a negative situation for Italy, whose dysfunctions have dramatic multipliers: like the other peripheries in the Mediterranean, Italy is subject to higher freight costs for being served second-hand by feeder ships, but it also suffers higher costs of land transportation because of the disorganized way in which the vast territory is served. Containers handled in the ports of Northern Italy are estimated to travel an average of 250 km [4], versus the 50 km of those moved in most ports of Northern Europe; the short distance doesn't justify train transport and leads to the almost exclusive use of trailers: that is 1.25 billion km a year, which can throw any motorway network into crisis.

But the dysfunction does not end here, because a single economic area, as is Northern Italy, that does not have a standard port with "drawing power" tends to develop piecemeal, multiplying the kilometres not only of the first journey from port to user, but also between the various stages of manufacturing, estimated by the confederation of Italian industries to be an average of 7 subsequent processes. Without any innovative solution, Northern Italy is therefore doomed to marginalization because of the higher costs of freight and land transport on the one hand and progressive paralysis due to the saturation of the road infrastructure, on the other.

### 3.2.4 Competition

Maritime transport is one of the industries with heavy competition, or rather is commonly considered a sector with perfect competition; small variations in demand can cause big variations in price with vicious circles that multiply the effects. The same is true for liner services, where competition is very strong and where we are not dealing within the field of the goods producing industries that compensate fluctuations with warehousing, but in that of the service industry, where any empty space at a ship's departure is lost space [3]. Shipping companies always have engaged in protective practices, uniting in conferences to keep freight costs low and limit competition. Antitrust tends to contest these practices, but the container has anyhow made it difficult to apply them, since the single, homogeneous box hides the identity of the individual items of cargo inside it. So far, the shipping companies have mainly aimed at forming consortia that bring together various groups and seek to integrate services in order to reduce costs and open up more opportunities; this practice, however, cannot exclude the competitors and offers the shipping companies rather few benefits.

Competition has further increased in recent years with the emergence of a new phenomenon in the shipping industry that is leading to a destructive type of competition. In fact, in the past the sector operated at a fairly constant level of technology and with moderate energy costs, while ship construction costs increased over the years due to inflation. As a result, new shipping companies struggled to remain competitive with the existing ones, which could use comparable ships that had cost less and were partially amortised. The popular saying was that a ship could pay off only after a prior bankruptcy: this means that a new ship was too expensive to be competitive and was only affordable after a bankruptcy, when it could be purchased at half price. This obviously presented a barrier to entry for new shipping companies.

Now we have the reverse situation: energy costs have increased dramatically, also because of the increased speed of ships, and exceed the total time charter costs; technological evolution has brought an explosion of solutions for ships that are more efficient thanks to their dimension and design and also consume less power because of their slim, streamlined shape and their high-efficiency engines. In addition to this, the current crisis in the industry has now also lead to a reduction of the construction costs.

In this situation, a new ship is cheaper than one that is only a few years old, because it has better features and less pollution problems, needs to pay off lower costs and consumes much less fuel. Someone has actually calculated (maybe using slightly exaggerated values) that an oil tanker built before 2007 (therefore almost new considering that a ship's service life can last more than 30 years) cannot be competitive even if chartered for free, because its higher fuel consumption costs more than the time charter (crew, amortisation, insurance and maintenance) of a new ship.

So the barrier to entry has fallen, creating a strong incentive to enter the market and thereby triggering a competition mechanism that is destructive. For the moment it is difficult to evaluate the consequences of this new phenomenon, but it is not easy to imagine how a service-producing business can survive in a situation of perfect competition, where a new entrant has the licence to kill those already there.

We are presently witnessing a crisis in the container transport market due to excess supply, especially for the Europe—Far East connections, where the major players continue to order new ships, not because they believe demand will increase and saturate the increase in supply, as claimed in many conventions, but because they are stronger and they know that by doing so, they can achieve a competitive advantage and eliminate weaker competitors.

This is, however, a self destructive practice, because it inevitably takes a long time to achieve this goal and in the meantime others could be tempted to enter the game and take advantage of the competitive situation, creating a vicious circle of forever operating in the red. Even if profit has reappeared on some balance sheets today, the margins are too limited and risky.

It seems as though the big players were aware of the inherent risks of a possible destructive competition scheme and in reply, breaking through the cultural and economic obstacles that hinder alliances, have planned a new alliance "P3" between the top three global shipping companies, Maersk (Netherlands), M.S.C. (Switzerland) and C.M.A. C.G.M. (France). As of the middle of next year they expect to launch joint services throughout the world. This hitherto unthinkable organisation will operate a fleet of 248 ships with a total capacity of 2.7 million TEUs and an average cargo carrying capacity of 11,000 TEUs per ship. Unless the

Antitrust intervenes, the pendulum of evolution of course might have gone too far in the other direction: this attempt to prevent destructive competition might not only eliminate excess competition, but the very competition itself, creating a *de facto* monopoly of global dimension that could do a disservice to global trade and seriously harm the production development and international integration.

Just as excessive competition is destructive, so is the creation of a monopoly that wants to fight it. The fact remains that it is not easy to find a balanced solution that protects both producers and users in everyone's interest. The coming years will show us whether a solution to this not so easy problem will have been found.

## References

1. Lami IM (ed) (2007) Genova: il porto oltre l'Appennino. Ipotesi di sviluppo del nodo portuale, Celid, Torino
2. Lami IM, Beccuti B (2010) Evaluation of a project for the radical transformation of the port of Genoa–Italy: sccording to community impact evaluation (CIE). Manage Environ Qual 21:58–77
3. Musso B (2008) Il porto di Genova. La storia, i privilegi, la politica. Celid, Torino
4. Musso B, Roscelli R, Lami IM, Rosa A (eds) (2009) Il BRUCO-Bi level rail underpass for container operations. Celid, Torino

# Part II
# Theoretical Contribution: Integrated Strategies, Governance and Decision—Making Processes

# Sectoral Drawbacks in Transport: Towards a New Analytical Framework on European Transport Corridors

4

Patrick Witte and Tejo Spit

**Abstract**

The present understanding of bottlenecks in the European transport network fails to grasp the cumulating and culminating effects of bottlenecks, for the scope of the research is in most cases limited to a one-sided (logistics) perspective. A theoretical framework has been created, which argues that bottlenecks should be interpreted as integrative, complex problems, operating on the cutting edge between transportation, spatial planning, environmental issues, economic development and transnational governance. This chapter will provide empirical evidence to support this framework, in a context of European transport corridor development. The theoretical framework has been tested in an empirical setting by zooming in on the European transport Corridor 24. In a first step, both general (macro-level) and specific (micro-level) bottlenecks have been identified by interviewing logistics experts. In a next step, these first results will be further used to perform an in-depth, qualitative analysis of bottlenecks in case-study areas along Corridor 24. One of the key findings is that bottlenecks emerge from different, sectoral perspectives. Moreover, these perspectives appear to be highly interrelated. In other words, more attention

This chapter is an adaptation of a research paper by the same authors, originally published in Research in Transportation Business and Management (2012, volume 5, pp. 57–66). This chapter aims to summarise the main argument of that paper, and add some new empirical insights obtained since the original publication.

P. Witte (✉) · T. Spit
Utrecht University, Heidelberglaan 2, 3508 TC, Utrecht, The Netherlands
e-mail: p.a.witte@uu.nl

T. Spit
e-mail: t.j.m.spit@uu.nl

I. M. Lami (ed.), *Analytical Decision-Making Methods for Evaluating Sustainable Transport in European Corridors*, Sxi 11, DOI: 10.1007/978-3-319-04786-7_4,

should be paid to the cumulating and culminating effects of bottlenecks, operating as comprehensive problem areas. The most important implication is that, when using a limited, sectoral perspective on bottlenecks, one loses track of the possible added value of sector-transcendent analyses. This will ultimately lead to inefficient use of transport networks. This chapter provides a new conceptualisation for the possibilities of inter-sectoral coordination in dealing with bottlenecks in the European transport network.

## 4.1 Introduction

The existence of bottlenecks in the European transport network is a persistent issue in European (spatial) policy. Therefore, the possibility of upgrading the existing transport infrastructure to help remove bottlenecks has been extensively studied in recent years. However, the upgrading of existing infrastructure is only part of the solution. And a lack of capacity in the infrastructure network—the reason for upgrading the infrastructures—is only part of the problem. This chapter argues that a lack of understanding of the scope, complexity and cumulative effects of bottlenecks is the most prominent aspect currently missing in the analysis of bottlenecks in the European transportation network.

The traditional understanding of transport bottlenecks is predominantly limited to a (technical or managerial) sectoral perspective of particular concern within this understanding are the capacity constraints of transport infrastructure. The technical capacity of transport infrastructure can be defined as follows, adapting Rothengatter's [17] definition of the theoretical capacity of a rail network: "*The maximum quantity of freight which can be operated on a link, depending on a number of factors such as the type of vehicles, the speeds, the mix of transport modes as well as the operation and scheduling systems*" (p. 51). This closely relates to a literal definition of a bottleneck as a narrow section of road or a junction that impedes traffic flow.

Despite the attention to transport bottlenecks, academic research thus far has largely failed to develop a comprehensive, consistent and especially an integrative framework to analyse and evaluate bottlenecks in transport networks. The urgency of resolving bottlenecks in European transport networks has heightened the need for innovative solutions. However, as will be pointed out in this chapter, this is easier said than done, since transport bottlenecks have become so much interrelated with a multitude of economic, spatial and governance issues. This has thus far only been partly understood. The aim of this chapter is therefore to shed more light on the complexity of the sectoral bottlenecks and their development into comprehensive problem areas in which the problematic characteristics of old (sectoral) and new (comprehensive) bottlenecks cumulate and culminate.

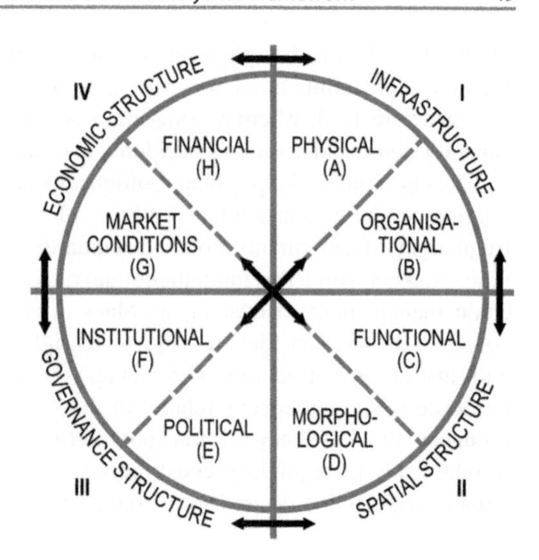


**Fig. 4.1** Conceptual framework (Reproduced with permission from [20])

The next section will discuss the various theoretical perspectives regarding bottlenecks. Next, the methodology used in the research will be presented. The empirical section highlights the key findings stemming from the various empirical studies that have been carried out. Finally, the main conclusions will be discussed.

## 4.2 Analytical Framework

This section will discuss different perspectives on transport bottlenecks. An extensive literature review has been performed. This has generated a conceptualisation consisting of four common, distinctive perspectives on bottlenecks (Fig. 4.1). The first is infrastructure (I), including the physical (A) and organisational (B) dimension. The second is spatial structure (II), consisting of the functional (C) and morphological (D) structure. The third is governance structure (III), dealing with the political (E) and institutional (F) structure. The fourth is economic structure (IV), taking into account the market conditions (G) and financial aspects (H). Within each perspective and type, numerous bottlenecks can be found. This chapter will highlight the most important ones.

### 4.2.1 Infrastructure (I)

In this chapter, the subdivision in physical and organisational structure which is common in definitions of infrastructure will be used to explain the different bottlenecks involved in the infrastructure perspective. First, physical bottlenecks (A) will be discussed. The most common bottleneck within this category is congestion. Congestion involves many dimensions, various spatial scales and multiple transport modes [3, 14]. Congestion should not be confused with another important type of bottleneck in the physical transport infrastructure: capacity constraints. Capacity constraints amount to the mere technical capacity of a certain piece of infrastructure [17], whereas congestion also originates from other issues besides capacity constraints, such as accidents and bad weather.

Closely related to physical bottlenecks are organisational bottlenecks (B), relating to the organisational *facilities* of infrastructure. Apparently, there is a frequent call for harmonisation and standardisation, originating from, for instance, policymakers who try to implement innovative concepts such as 'integrated supply chain management'. However, as Maes et al. [9] point out, there is hardly any cooperation between logistics and industrial companies. It proves hard to break with institutional structures. The problems with harmonisation and standardisation reinforce other bottlenecks related to the organisational infrastructure, which all influence the efficiency of transport networks: for instance, the adaptation of freight loads to regulatory constraints [15]. It goes without saying that a more holistic approach is desired to overcome these problems.

### 4.2.2 Spatial Structure (II)

The second perspective concerns the spatial structure of transport networks, consisting of the functional and morphological structure. The functional structure covers aspects related to land use, plus the planning processes underlying the actual land use. The morphological structure covers the unplanned, external conditions or surroundings, especially those in which people live or work. First, bottlenecks related to the functional structure (C) will be discussed. Actual land-use bottlenecks can be summarised as 'pressure of space on the transport network' [7]. One of the main issues is the lack of land for expansion in traditional port areas. This leads to changing port–city relations and expansion of ports towards the coast [19]. Bottlenecks related to the planning process are especially difficulties of involving private parties in the financing of transport infrastructure. Issues in this case are the diversity of actors and the risk-avoiding behaviour of private parties. Other constraints relate to multiple ownership of land or fragmented land ownership (e.g. [8]. Issues in this case are the behavioural characteristics of land owners and the institutional context of land ownership.

With regard to the morphological structure (D), two types of bottlenecks emerge. The first type can be characterised as traffic externalities, in most cases implying environmental effects. The externalities consist of the degradation of urban landscapes, use of space by traffic, road safety (i.e. accidents), air pollution and other types of environmental pollution, traffic noise, etc. [1]. The second type of morphological bottleneck concerns 'inescapable' physical barriers, in which path-dependent development has a crucial part to play. An example is the passage through the Alps to reach the seaport of Genoa in Italy, by means of the hinterland connections of the Port of Rotterdam; tunnels are still being constructed and the topography does not allow for very high speeds [18].

### 4.2.3 Governance Structure (III)

The third perspective is related to the governance structure. Governance structure can be divided into political structure and institutional structure. With regard to the political structure (E), different bottlenecks emerge. A first issue is the lack of knowledge of politicians and the subsequent use of planning methodology in practice. As Peters [12] suggests, European Union (EU) transport investments lack consistency and sustainability owing to the existence of partially complementary, partially competing development objectives. Furthermore, planning processes especially in transport corridors are often characterised by a narrow focus on bottlenecks and a rather defensive attitude taken by regional and local governments [16].

The second dimension of the governance structure is the institutional structure (F). This type closely relates to the organisational bottlenecks mentioned before. In this chapter, organisational bottlenecks concern friction factors with regard to the organisational *facilities* in infrastructure (formal, hard structures). Institutional bottlenecks are defined broader and thus cover also the way people (or firms, public bodies, etc.) *make use* of these facilities (informal, soft structures). Institutional fragmentation can be regarded as an example of a serious institutional bottleneck. Institutional fragmentation occurs in situations where different procedures do not fit with each other. This is the case in the European rail system, for instance, where a host of different technical systems is used by national rail companies simultaneously [13]. A related issue of fragmentation is the transnational nature of transport corridors, cutting through regional and national administrative borders [16].

### 4.2.4 Economic Structure (IV)

The final perspective is the economic structure. The definition of economic structure is the availability, quality, spatial distribution and cohesion of production functions, including infrastructures. To be more specific, economic structure will be divided into market factors (i.e. conditions) and financial factors (i.e. availability and allocation of resources). Bottlenecks related to market conditions (G) can be characterised as the influence of competition and market principles on the one hand, and the effects of agglomeration externalities on the other. In the first case, one can point to operational and commercial barriers obstructing access to infrastructure. Another example is the existence of monopolistic structures in transportation networks. Regarding agglomeration externalities, bottlenecks that can be identified are to be found at the 'break-even point' where positive agglomeration effects turn into negative agglomeration effects. There are limits on the degree to which agglomeration contributes to economic growth, particularly in metropolitan areas, where congestion and environmental degradation can become important problems when this 'turning point' is reached [6].

Bottlenecks related to financial factors (H) consist of both the basic availability of financial resources and the costs and effects of the actual allocation of these resources. Concerning the availability of financial resources, one should not be surprised that the recent economic downturn is regarded by some researchers as an external factor which is disturbing and damaging the already declining funding activities of governmental bodies. If investing in transport infrastructure occurs nevertheless, there are oftentimes many problems. Examples include the costs of investments [10], diverse effects of over- or under-building of infrastructure [11] and the unlikelihood of short-term returns on infrastructure investments [18].

### 4.2.5 Cumulative Effects of Bottlenecks

In many cases bottlenecks appear to be interrelated, leading to cumulative effects. This links closely to the concept of logistical friction as a multidimensional concept [7]. Friction factors can be understood as factors impeding the (most efficient) circulation of freight. The different perspectives of bottlenecks therefore essentially are friction factors, which cumulate to create a bottleneck. The strongest relations can be found between the infrastructural perspective on bottlenecks and any of the additional perspectives on bottlenecks. Infrastructure and spatial structure are connected for instance by the integration of transport infrastructure in the urban fabric and local environments. The negative external effects created by traffic externalities are another example. The pressure of space on transport, for example through the effects that the operation of real estate markets has, is also illustrative. Finally, the negative impacts of environmental protection on the transportation network can be mentioned.

The relation between infrastructure and economic structure has also been pointed to before. One can think of the friction between policy documents aiming at the introduction of new concepts such as integrated supply chain management, the competitive considerations of logistics companies and the financial consequences that could possibly follow a decision to implement such concepts. This also links closely to the relations existing between infrastructure and governance structure. In this case, the correlation between technical and organisational chokepoints (electric power compatibility, waiting times, interoperability) and the political and institutional embeddedness of these chokepoints comes to the fore.

On basis of the foregoing, an integrative conceptual framework is designed (Fig. 4.1). This conceptual framework can be understood as a wheel, which consists of a number of different spokes. The wheel consists of four quadrants (i.e. the perspectives), and each quadrant of two types (i.e. the dimensions). Returning to the concept of logistical friction, it should be stressed that there are many different friction factors hampering the most efficient movement of freight. Each spoke in the wheel can therefore be understood as a friction factor cumulating to create a bottleneck. The main argument here is that in attempting to solve a bottleneck, it is not sufficient to consider only one dimension. Because of the cumulative effects of bottlenecks, all types of friction factors should be considered. The arrows in the model represent the connectedness of all the different perspectives involved. There is no specific order in arranging the quadrants in the model, nor in the length or magnitude of the arrows. The model is used merely as a visualisation of the complex overlaps of bottlenecks.

## 4.3 Methodology

The conceptual framework will be tested in an empirical setting by zooming in on the TEN-T Corridor 24 transportation network. Corridor 24 is one of the major transport corridors in north-western Europe, stretching from Rotterdam to Genoa. Transport corridors can be defined, following Priemus and Zonneveld [13], as bundles of infrastructure (roads, railways, waterways) connecting two or more urban regions (p. 167). Transport corridors are concerned with connections (i.e. transport nodes) that use different modes (road, rail, barge or intermodal) and include both passenger and freight transport. This empirical application deals with the European area interested in the development of the intermodal transport corridor linking the transport nodes of Rotterdam and Genoa. This space hosts a number of the most densely populated urban regions in Europe (Fig. 4.2). What becomes evident is that different spatial scales are at stake on the Corridor 24 transport network. The transnational transport corridor scale (macro), as well as the urban region and the local transport node scale (micro), are of importance. Therefore this chapter needs a methodology that is suitable for both the macro and the micro level of analysis at the same time, since neither of the two levels is able to capture the full complexity of the transport bottlenecks occurring on transport corridors.

This chapter will use mixed scanning methodology, since the requirements mentioned above especially coincide with the overall principle of this type of methodology: a broad, strategic analysis of the main bottlenecks on the transport corridor will provide a definition of specific problem areas which require a more detailed examination. Mixed scanning methodology was originally introduced by Etzioni [4, 5] and is still often used in decision-making and planning. Etzioni used the metaphor of a weather observation system to explain the logic and relevance of his framework. Where a rationalist would examine the entire sky and an incrementalist would focus on certain, specific areas, the mixed scanning approach uses two cameras: a broad angle to cover all parts of the sky, but not in great detail, and a second camera to focus on those areas revealed by the first camera to require a more in-depth examination [4]. When applying the mixed scanning framework to this chapter, one could argue that a broad, strategic analysis of the transport capacity of the transport corridor will fail to take into account specific, interrelated chokepoints at the local level. For example, the impact of noise protection measures resulting from national legal structures. At the same time, a regional strategy for a certain transport node will neglect the impacts of border-crossing problems on the transport corridor at the transnational level. Mixed scanning is able to tackle these problems.

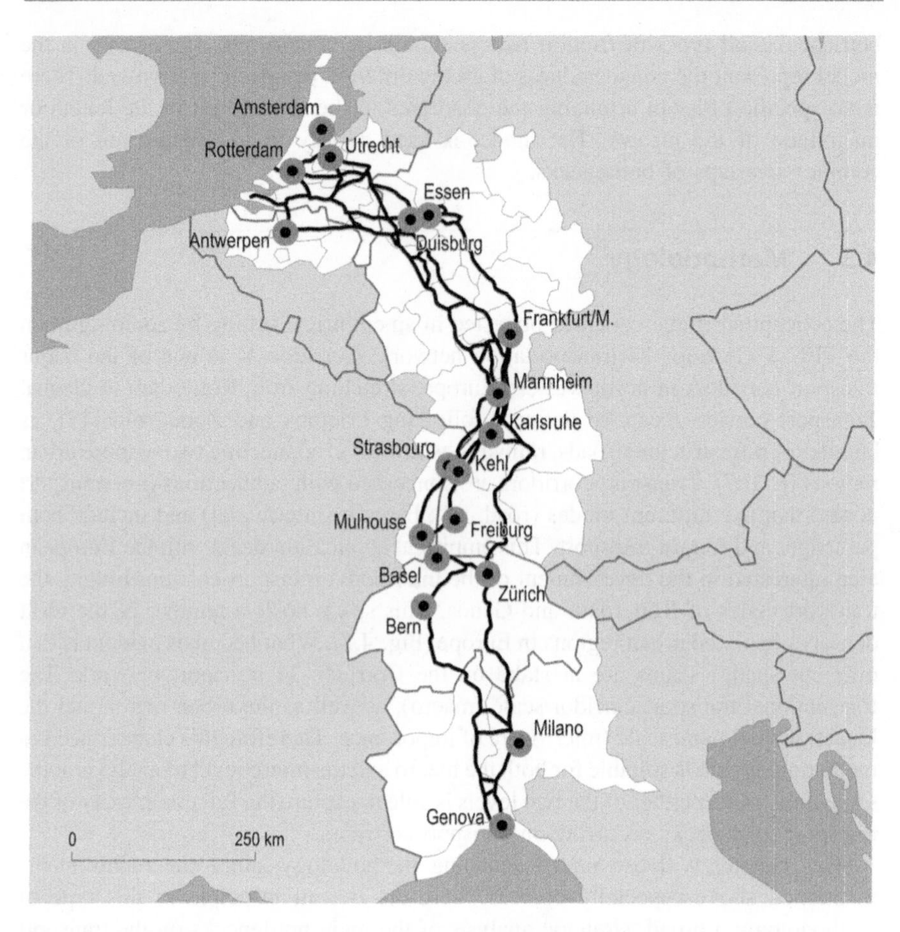


**Fig. 4.2** Corridor 24 and its environment (Reproduced with permission from [20])

The identification of general (macro) bottlenecks for the Corridor 24 regions has already been described extensively in Witte et al. [20] and will therefore not be repeated at this point. This chapter only makes use of data derived from the Regional Workshops, which provided this chapter with a selection of problems in specific locations (micro) along the Corridor 24 transport network[1]. The aim of these workshops is to be open to all the institutions and citizens interested in the Corridor 24 transportation network by activating a network of strategic decision makers and stakeholders and starting up a series of workshops to share information and collect expectations on a regional (micro) scale. In this way nine Regional

[1] The research has been carried out in the context of the INTERREG IV-B co-funded project CODE24.

Workshops have been carried out.[2] About three hundred people, including regional and local planning authorities, transport authorities, logistic and transport entrepreneurs, research institutes and experts, local companies and global corporations, associations of citizens, port authorities and political decision makers, participated. In addition, this chapter will highlight some findings from an informal test planning procedure, which was undertaken in the region of Wesel, Germany. These results can be viewed as a first attempt to perform an in-depth, qualitative analysis of bottlenecks in case-study areas along Corridor 24. It needs to be stressed that these results can only be used to gain a first, indicative impression of the empirical validity of the conceptual framework. Follow-up case study evidence will be the topic of debate in another, forthcoming paper.[3]

## 4.4 Results

This section will first summarise the micro-level findings of Witte et al. [20]. These findings have resulted from the Regional Workshop in Rotterdam, complemented with follow-up in-depth interviews with logistics experts from the Port of Rotterdam Authority. Nine experts participated in the Regional Workshop, including representatives of the Dutch Ministry of Transport, the Port of Rotterdam Authority, public and private institutions in the management of Dutch railway systems and universities. Afterwards, the indicative case study evidence from the Wesel region in Germany will be presented.

### 4.4.1 Micro-Level Analysis

One of the key findings of the micro-level analysis is that a customer perspective, which stresses the need to perceive bottlenecks from the point of view of direct users of transport infrastructure, is the most prominent aspect lacking in the present understanding of bottlenecks. This is reflected in a number of technical and managerial bottlenecks (Table 4.1). The lack of a customer perspective plays a key role in the discussion of all the bottlenecks identified in the micro-level analysis.

In the Dutch context, a number of issues related to infrastructure (I) were identified. With regard to the physical (A) point of view, issues identified were a lack of long tracks at the starting points of freight routes, a lack of sufficient capacity along the way (i.e. too many trains are operating on the same tracks), a lack of long tracks at the train stops along the way and too many different systems. From a customer point of view, the previous physical problems result in several

[2] The workshops took place in Rotterdam (Netherlands), Antwerp (Belgium), Essen, Frankfurt, Mannheim and Karlsruhe (Germany), Zurich (Switzerland) and Milan and Genoa (Italy), thus covering the entire space belonging to Corridor 24. The authors participated in the workshop in Rotterdam.

[3] This paper will deal with bottlenecks in a context of inland ports development.

**Table 4.1** Technical and managerial bottlenecks in the Netherlands

| Technical | Managerial |
|---|---|
| Track length | Needless stops |
| Track capacity | Travel time |
| Train length | Circulation time |
| Security systems | Estimated time of arrival |
| Voltage systems | Knowledge of trains' priorities |
| Slot incompatibility | Traffic management |
| Free access to ports | Cross border slot reassignment |
| Connections to terminals | Language barriers engine's drivers |

(*Source* authors' own table based on expert interviewing)

organisational (B) problems: transporters cannot operate trains with a length of over 700 m, they need very expensive engines regardless of the distance travelled, they have to make needless stops and they cannot make ideal circulations because of timetables and working conditions of engine drivers.

Related to bottlenecks in infrastructure are bottlenecks in spatial structure (II). The morphological structure (D) is of special interest in this case. It appears that many present-day bottlenecks result from past path-dependent choices that are reflected in the present spatial, morphological structure. Examples of specific bottlenecks in the Netherlands include different security systems along the A15 highway, 1,500-volt 'islands' (compared to 25 kV continuous-flow electricity systems), too short tracks on Maasvlakte–Oost and Waalhaven (Rotterdam), limited transport capacity 'at the doorstep' and a lack of tuning between limited slot-capacity and the ideal of an accurate 'estimated time of arrival'. In part these bottlenecks can be considered as 'accessibility problems' in traditional port areas. A lack of accessibility can also be characterised as a bottleneck in the functional structure (C).

When extending the analysis to include a cross-border corridor perspective, bottlenecks in infrastructure are complemented by bottlenecks in governance structure (III). Some experts mentioned the problems identified in the NewOpera report. Those include insufficient cross-border coordination for slot reassignment; a lack of harmonisation in train numbering, tracing and handling; a lack of supporting tools to manage traffic; a lack of knowledge of trains' priorities; and a lack of punctuality [2]. The key finding of this research report is that technical improvements on the corridors will be nullified if driving rules, working patterns and safety regulations are not standardised.

Of course, there are many programmes and actions going on to tackle these problems. However, as the experts have repeatedly stressed, as long as 'the customer' does not take part in these projects, effects will be small. A promising solution would be to classify and deliver programmes and actions according to the customer's preferences. This is, however, easier said than done; projects often diverge and there are no strict deadlines for realisation of such projects. This issue is closely related to the political (E) and institutional (F) bottlenecks.

The different perspectives (in brackets) again seem highly interrelated, leading to the cumulative effects of bottlenecks. For example, to upgrade the present level of service in railway freight transport operations (organisational), several measures are needed (e.g. improvement of reliability, shorter travel times). What is required to achieve these improvements is, for instance, an attitude shift from reactive to proactive on the part of the infrastructure managers, railway undertakings and terminal operators (institutional) and a close cooperation between various traffic managers (market conditions). Besides, heavy investments are required (financial) to further improve the functioning of the present transport infrastructure network (physical). Examples include the implementation of the ERTMS security system at the Kijfhoek shunting yard (near Rotterdam, Netherlands) or near the Zevenaar border (close to Emmerich, Germany).

But who will pay? The experts have agreed that there is a need for an 'integrated corridor director' to mediate in such issues. Ideally, 'the market' should initiate such a director, but in certain cases, the experts concluded, 'the market' also profits from suboptimal solutions. There appears to be a lack of involvement; no one is willing to invest. This is a clear example of the effects of economic structure (IV) on the Corridor 24 transport network. In summary, one can state that 'the customer' is often lacking in discussions on bottlenecks, especially in corridors, and that an integrated corridor director is suggested as a promising way forward.

### 4.4.2 Informal Test Planning Procedure: The Case of Wesel, Germany

A recent example of the integrated nature of bottlenecks in European corridor development is the discussion with respect to the creation of a third railway track in Germany between Emmerich and Oberhausen to better connect the dedicated Dutch freight transport railway line 'Betuweroute' to the German hinterland. This railway line has a strategic importance as a freight corridor connecting the Port of Rotterdam to the Ruhr region in Germany. Whereas the Dutch government has speeded up the procedure for implementation of this project, the German procedure is running parallel, but without strict deadlines for implementation, owing to (national) political reasons. This is likely to hamper the implementation of fluent cross-border freight transport in the short term.

The German rail operator 'Deutsche Bahn' has therefore developed a project to upgrade the railway line to three tracks and eliminate most of the crossings. According to the German law the new development should provide the necessary compensations to the communities, including noise reduction measures (i.e. noise walls). However, due to the topographical structure of the area and the type of settlements, these walls need to be high (often between 2 and 6 m). This has encountered the opposition of the communities that see the proposed solution as a further disturbance to their living condition rather than betterment.

At first sight, this seems to be merely a transportation bottleneck; there is lacking capacity on the German part of the network following the Betuweroute, so an additional railway track is needed at one specific section of the network. However, closer examination also reveals problems with respect to transnational governance: political resistance to the project, and differences in institutional structures and procedures, which hamper efficient cross-border cooperation. Moreover, the German section of the line presents several additional problems that need to be solved: some fifty level crossings along the line, insufficient capacity of the stations (e.g. Oberhausen) or sub-optimal employment of the nodes and disturbances to the surrounding settlements (e.g. noise, dangerous materials, fragmentation of the communities).

First of all, it should be noted that many problems are interrelated in this case. What at first sight seems to be a mere transportation bottleneck also appears to have clear spatial, environmental, political and institutional dimensions. In addition, the issues occur on multiple levels of scale. On a local level, the project of the Deutsche Bahn is facing heavy local resistance because of the visual impact of the noise walls. On the other hand, from a transnational corridor perspective, this area is of crucial importance to achieve efficient goods travel from the Betuweroute to the German hinterland. Thus, different issues interfere at different spatial scales. This calls for a strategic set of measures.

A second point of concern is the modal competition this area is facing. On the one hand, the creation of a third track to solve the problems should be measured against the alternative costs of expanding the German motorway network. The recent policy attitude towards achieving modal shift from transport by road to rail and inland navigation is helpful and strategic in this respect, to strengthen the insufficient and difficult links between the railway and the inland ports in this region. On the other hand, inland navigation itself via the river Rhine can also be seen as a competitor to rail transport for this area. Moreover, this line is in competition with other corridor routes that also show high rates of ton/km and with other projects that also opt for German federal funding. In this way, economic potential can also be included as factor of importance, to add to the complexity of this area.

To contribute to a solution, an informal test planning procedure called '*Ideenwerkstatt Fortsetzung der Betuweroute*' has been promoted by the regional association Ruhr together with the municipalities along the German part of the Betuweroute. The aim of this informal procedure is to elaborate alternatives to high noise barriers that separate entire settlements as a foreign body structure and to find alternative, innovative and original solutions. Three cities in the region 'Kreis Wesel', Dinslaken, Wesel and Hamminkeln (Mehrhoog), were selected as pilot areas where these attempts should take place.

Some strategic questions can be posed regarding this region. First, should this project concentrate only on the creation of a third track and the realisation of noise protection on a local scale to solve the bottlenecks, or are there more strategic interventions to be implemented in this region? Second, is noise protection the only way to tackle the existing environmental bottlenecks, or can the upgrading of the railway station areas and their surroundings lead to synergies on a regional scale, which can be seen as a form of compensation?

In the first question, the negative external effects which tend to aggregate on a low spatial scale (i.e. noise nuisance, safety and visual quality of the localities involved) are measured against possible positive spill-over effects which tend to aggregate on a higher spatial scale (e.g. corridor development and related regional economic growth). The second question makes use of a growth management perspective: in this case compensating noise nuisance with the creation of synergies at railway station areas. In this way, transport, spatial and environmental issues on a local scale are tried to be solved by seeking economic potential on a regional to transnational (corridor) scale. The research problem thus evolves from a short-term technical transportation issue to a problem of long-term economic development and possible planning interventions.

In summary, the strategic questions that were posed with regard to the corridor issues in the region of Wesel have contributed to the development of a regional and integrated perspective on the future development of the region. By focussing so strongly on the local noise problem, the regional development perspective had been lost. This is not to say that technical solutions to technical bottlenecks are therefore irrelevant, but by adopting this integrated perspective new opportunities and development alternatives have come to the fore.

## 4.5 Conclusions

Intermodal transportation is often hampered by bottlenecks in transportation networks. So far, the understanding of these problems has remained largely incomplete. Policy documentation is often limited to include only sectoral perspectives on bottlenecks. Especially in times of economic downturn, a sectoral perspective is often favoured over a holistic approach towards bottlenecks, for reasons of efficiency. One can think here of the traditional emphasis on the literal definition of bottlenecks, that is, the mere capacity constraints and congestion occurring in the infrastructural networks.

What has become clear, however, is that bottlenecks can no longer be viewed as mere capacity constraints of infrastructure networks. Instead, they should be interpreted as being integrative, complex problems, operating on the cutting edge between transportation, spatial planning, environmental issues, economic development and transnational governance. In other words, more attention should be paid—both in scholarship and in practice—to the cumulating and culminating (friction) effects of bottlenecks, operating as comprehensive problem areas.

This chapter has suggested that bottlenecks emerge from different sectoral perspectives. Moreover, these perspectives are highly interrelated. Based on these suggestions, a conceptual framework has been developed to identify and analyse bottlenecks in a more holistic way. This can be considered a useful tool to the further development of planning education on this topic. The most important insight for practitioners in applying this framework is that when using a limited, sectoral perspective on bottlenecks one loses track of the possible added value of sector-transcendent analyses. This will ultimately lead to inefficient use of transportation networks, as the case of Wesel has demonstrated.

A suggestion to enrich the planning education on bottlenecks might be to rate the (lack of) importance of different types of bottlenecks as perceived by the direct users of transport infrastructure (logistics companies, port authorities, other relevant stakeholders, etc.). In this way it would be possible to arrive at a better understanding of the relative value of bottlenecks (i.e. the distribution of the fields in the model and the direction and magnitude of the arrows). This would also be an interesting way of asking private companies valuable information on bottlenecks without having to ask them for sensitive data or information. This method could result in a clear and easily interpretable framework for practitioners to deal with comprehensive bottlenecks in the European transportation network.

## References

1. Banister D (2000) Sustainable urban development and transport—a Eurovision for 2020. Transp Rev 20(1):113–130
2. Castagnetti F (ed) (2007) NewOpera—the rail freight dedicated lines concept. The European Freight and Logistics Leaders Forum, Brussels
3. Chapman D, Pratt D, Larkham P, Dickins I (2003) Concepts and definitions of corridors: evidence from England's Midlands. J Transp Geogr 11(3):179–191
4. Etzioni A (1967) Mixed-scanning: a 'third' approach to decision-making. Public Adm Rev 27(5):385–392
5. Etzioni A (1986) Mixed scanning revisited. Public Adm Rev 46(1):8–14
6. Farole T, Rodriguez-Pose A, Storper M (2009) Cohesion policy in the european union: growth, geography, institutions. London School of Economics, London
7. Hesse M, Rodrigue JP (2004) The transport geography of logistics and freight distribution. J Transp Geogr 12:171–184
8. Louw E (2008) Land assembly for urban transformation—the case of's hertogenbosch in the Netherlands. Land Use Policy 25:69–80
9. Maes J, van de Voorde E, Vanelslander T (2009) Mapping bottlenecks in the flemish logistics sector. European Transport Conference 2009
10. Marvin S, Guy S (1997) Infrastructure provision, development processes and the co-production of environmental value. Urban Stud 34(12):2023–2036
11. McCann P, Shefer D (2004) Location, agglomeration and infrastructure. Pap Reg Sci 83:177–196
12. Peters D (2003) Cohesion, polycentricity, missing links and bottlenecks: conflicting spatial storylines for Pan-European transport investments. Eur Plan Stud 11(3):317–339
13. Priemus H, Zonneveld W (2003) What are the corridors and what are the issues? J Transp Geogr 11:167–177
14. Rodrigue JP (2004) Freight, gateways and mega-urban regions: the logistical integration of the bostwash corridor. Tijdschrift voor Economische en Sociale Geografie 95(2):147–161
15. Rodrigue JP, Debrie J, Fremont J, Gouvernal E (2010) Functions and actors of inland ports: European and North-American dynamics. J Transp Geogr 18:519–529

16. Romein A, Trip JJ, de Vries J (2003) The multi-scalar complexity of infrastructure planning: evidence from the Dutch-Flemish megacorridor. J Transp Geogr 11(3):205–213
17. Rothengatter W (1996) Bottlenecks in European transport infrastructure. In: Proceedings of the 24th European transport forum 401:51–77
18. Van Klink HA, van den Berg GC (1998) Gateways and intermodalism. J Transp Geogr 6(1):1–9
19. Wiegmans B, Louw E (2011) Changing port–city relations at Amsterdam: a new phase at the interface? J Transp Geogr 19:575–583
20. Witte P, Wiegmans B, van Oort F, Spit T (2012) Chokepoints in corridors: perspectives on bottlenecks in the European transport network. Res Transp Bus Manag 5:57–66

# Integrated Spatial and Infrastructural Development: The Need for Adequate Methods and Spatial Strategies for Collaborative Action and Decision-Making

5

Bernd Scholl

**Abstract**
In the Lisbon Agenda, Europe formulated the goal to become the most competitive economic region worldwide. The present economic crisis makes it clear though, this will be a stony path. For one, Europe doesn't have spatially related strategies and priorities to identify reliable spaces for future investments. And second, building up an efficient trans-European transport network must be a key element of this strategy. The primary element of a sustainable spatial and settlement development is a high-capacity railway network to serve as a kind of backbone. To take full advantage of the synergy, future settlement development should be concentrated in the catchment areas of the train stations of this high-capacity public transport. These two elements are critical to achieving this desired competitive advantage. There is a need for adequate methods and strategies to achieve this. The case of the Corridor Development 24 (CODE24) Rotterdam–Genoa illustrates the challenges, methods and strategies needed to achieve an integrated spatial and infrastructure development.

## 5.1 Introduction

In the Lisbon Agenda, Europe formulated the goal to become the most competitive economic region worldwide. The present economic crisis makes it clear though, this will be a stony path. For one, Europe doesn't have spatially related strategies

B. Scholl (✉)
Institute for Spatial Planning and Landscape Development,
ETH Zurich, Zurich, Switzerland
e-mail: bscholl@ethz.ch

I. M. Lami (ed.), *Analytical Decision-Making Methods for Evaluating Sustainable Transport in European Corridors*, Sxi 11, DOI: 10.1007/978-3-319-04786-7_5,

and priorities to identify reliable spaces for future investments. And second, building up an efficient trans-European transport network must be a key element of this strategy. The primary element of a sustainable spatial and settlement development is a high-capacity railway network to serve as a kind of backbone. To take full advantage of the synergy, future settlement development should be concentrated in the catchment areas of the train stations of this high-capacity public transport. These two elements are critical to achieving this desired competitive advantage.

Although Europe is fairly well advanced on this path and well positioned in comparison to other economic areas on other continents, there are still major challenges to be overcome. For example, for the railway system to be able to exploit its full potential, the missing connections must be completed and the bottlenecks in the transport infrastructure must be eliminated. In many places, bottlenecks in the infrastructure are hindering the further development of the corridors, e.g. Rotterdam to Genoa. Contradictory interests at various political levels result in conflicts that can restrict the ultimate development of the Corridor and the regions situated along it.

While the European Union is trying to promote the free transport of goods, capital, services and workforces by dismantling business and trade constraints, national egoism is leading to delays and setbacks, sometimes even a standstill, in the development of the priority corridors. However, there are also other barriers and constraints, so the communities often follow a strategy of redevelopment support for a local, efficient, high-capacity passenger public transport Hesse [2]. But, their goals are often thwarted or even made impossible by the transport of goods on existing routes that pass through densely settled centres, because the impact of freight traffic is counterproductive for a further concentration of housing development in the catchment areas of railway stations Therefore, the inhabitants affected by this situation have to be given even more consideration in the future expansion of the infrastructure. Spatial tolerance is therefore an important criterion for the construction of new lines or changes to the operation of existing lines. The present planning insecurity is a constraint and has negative effects and consequences in some quite different aspects.

Clarity in the joint infrastructure concept is not only important for transport development, it is also important for the development and operation of a future network of cities and sites and thus for spatial development. Therefore, developing a joint infrastructure concept should be considered an integral task of European importance.

This increases the importance of the supra-urban aspect of spatial planning, namely, regional planning as the middle level between national and community planning levels. Regional planning can, when it is action-oriented, bundle the wishes and demands of the local planning and translate them into a suitable orientation for regional development. This requires that the targeted direction is balanced with national planning development perspectives and with the plans of neighbouring regions in order to open and secure manoeuvrability for further development.

One remarkable occurrence is that it was the top regional level of the Regional Association of Rhine-Neckar that took the initiative to start a project of strategic European importance: the development of the corridor from Rotterdam to Genoa. It stands as an example that an overall strategy for regional cross-border planning needs exceptional efforts and many years of intensive cooperation. The actors of regional planning now, as before, are facing major challenges.

## 5.2 Indicators of Supra-Local Spatial Planning

In order to understand the challenges better, some of the important indicators of supra-local planning are listed here:

1. The size of the space and the number of important events and relationships, as well as the sheer amount of information, forces the use of abstraction; this is also about being able to separate the essential from the unessential. However, clarifying and solving spatially relevant problems also requires making them concrete. Just being able to connect one with the other is extraordinarily challenging. Abstraction requires (often radical) simplification, which brings with it the danger of falsifying some cases and their relationships, connections, while concretizing requires in-depth penetration, which carries the danger of getting lost in unimportant details. Therefore, an approach that uses *only* the abstract or *only* the concrete levels is completely unsuitable for complex, unclear problems.
2. The procedures that determine the spatial environment and its use are also complex. Many of these procedures do not allow themselves to be correctly represented through explicit relationships, e.g. laws of nature.
3. The number of actors participating in working out suggestions for evaluation and implementation is usually very large and requires special provisions for coordination, cooperation and communication.
4. Numerous important pieces of information that form planning's basic principles are data on quantitative distribution areas. For example, it could be that Switzerland will have 10 or 11 million inhabitants by 2050, and far more is also possible. Planning must be able to handle these uncertain predictions. In addition, some of the most important information could be founded on assumptions, opinions, points of view, judgements and prejudices.

These few indicators make it clear that supra-local spatial development and the actors of regional planning (especially) face a major methodological challenge to concentrate on tasks that cannot be resolved with the usual routines in a situation of consistently limited resources.

The rapid development of information technology has brought completely new possibilities with it. The more possibilities there are to generate and exchange information, the more challenging and important an effective approach will be. In the following sections, an appropriate approach will be sketched out. One essential goal is to build dams against the flood of information, while also ensuring that decision-relevant information is available at the right time.

## 5.3 Spatial Planning and Decisions

Spatial planning is a task-oriented discipline, and at the core, most problems can be traced back to problems of decision-making. The task is to clarify the main possibilities of spatially relevant decisions and actions.

What are possible basic options for action? What conditions should be observed? For example, if one assumes there are sufficient financial resources available for specific plans and then it turns out that no money has been reserved for them, then the reason for this oversight is always the first question. What side effects and consequences might arise as a result? Once an option has been selected, the question of how to implement it brings up operational questions. Who should do what with which resources and by when?

The answers to difficult problems, which, of course, are not limited to spatial planning, are not immediately obvious and require, at least in part, longer processes to work them out. Intended improvements are often implemented years and sometimes even decades later. Extensive infrastructure plans in the area of roads and railways, airport expansions and the transformation of settlement areas in urban and regional contexts often deliver clear descriptive examples. In Switzerland, for example, the Gotthard Base Tunnel, which is the core piece of the North-South Link from Rotterdam to Genoa, is anticipated to go into operation at the end of 2016, nearly twenty years after the start of construction. Whether intended effects, such as the extensive transfer of goods transport to the railways, can be achieved and whether undesirable effects, such as a displacement of regional transport, can be avoided cannot be foreseen.

Because the lead time, understood as the time between the actual selection of a specific option and the final results, in many spatial planning processes stretches out over longer time periods, functional procedures and methods take on increased importance. These should, and it's actually required in democratic nations, also include transparent and sustainable decisions and activities that contribute to the causative planning argumentation.

At their core, methods should support logical decision-making. The approach towards information gathering plays a critical role in such a process. One possibility is gathering information that *could* be important for the task by bringing possible solutions closer, while the other possibility finds out which information is important, and what is not, through attempts to solve the problem. Basically, the objective and subjective imperfection of the information should be carefully observed. Actors can only contribute partial information to the solution, based on their own available knowledge and skills. Whether a piece of information is important or not can be recognised if one can change the choice for a specific option and this changes the final decision. This so-called Maxim of Effectiveness, which was formulated in a very understandable way by the economists Modigliani and Cohen [3],[1] can have a considerable influence on the design of planning and processes. However, one gets the impression that this practical maxim has received too little attention up to this point.

It is obvious that a problem and its related decision-oriented approach allow an effective handling of information and based on that, problems at the starting point should be the basis for the selection and employment of the methodology. In addition, the first approximate ideas of possible solutions can help follow the above-mentioned maxim.

One of the common difficulties to master is avoiding piling up useless information for the problem clarification. Gathering and generating information often supports the illusion that more information leads to surer decisions. Sometimes, for tactical reasons, it is an attempt to confuse—or in any case that is the result.

Another outcome of collecting too much information is that, even considering that today time budgets are already limited, it seems that ever less time remains for finding an actual solution to the problem. Apparently, here, this has something to do with a thought pattern or paradigm that assumes that pulling all possible forms of problem description and clarification into the foreground is more important than putting the concentration required for spatial planning and development into the problem's solution. However, according to the philosopher Popper [6], it is not necessary to be more specific than the problem requires or as he put it: Clarity comes before unnecessary precision.

## 5.4 Consequences of the Applied Methodology

If one follows this approach, then information gathering procedures must also be adjusted to avoid superfluous, time-consuming and expensive data collection. The method should contribute to gradually penetrating the core of the difficult task, thus helping to differentiate important data or facts from the less important. When less information is needed, then the information is also less specific, and therefore less risky because it is not tied to a specific outcome. Or, from another perspective, decisions are more robust when they are flexible enough to deal with possible changes.

When less information is needed, communication is also lighter because the required mutual knowledge of the participants remains manageable. New information and communication technology can lead to a 'flash flood' of information and thus building dams against them is an important task. The uninhibited application of new technology can bring about the opposite from that which was actually intended.

[1] Original citation: 'Don't devote resources to estimating particular aspects of the future, no matter what you find out (with due consideration for what you might conceivably find out), if you would not be led to act differently from the way you would have acted without finding out'.

Specific demands on purposeful procedures emerge from the consequences and I want to go into two aspects in the following sections: the conscious approach to insecurity, risk and surprises and the importance of robustness for the decisions taken.

## 5.5 Approach to Insecurity, Risks and Surprises

Spatial planning decisions are always sensitive to insecurity issues. It can even be said that in this context no secure decision exists. What that means is that the intended effects of a decision may not be met or may be only partially met and undesired effects can be the result. Methods that offer apparent security should definitely be rejected. Jouvenel already suggested this in 1967: 'For the purpose of forecasting, however, one must arrive at various, preferably independent prognoses, which then mutually control each other' De Jouvenel [1]. Unsuitable methods, for example, are prognoses that only use historical observation developments as their media, then project them into the future and assign specific assumptions to the model-related simulation(s) as given hypotheses, postulates, or assumptions. Thus, the assumptions that may prove to be useful are as unknown as the solution itself.

It would it certainly be unreasonable, however, to accept more risks than those that are absolutely necessary under the given circumstances. The solution of difficult tasks requires therefore an understandable and systematic approach to risk, which should lead to avoiding unnecessary risks.

A further characteristic of the handling of complex tasks is the appearance of surprises. This is understood as the revelation of an experience or facts that were not previously known, something that was unexpected (things that you do not know and even did not know that you did not know and, above all, for which one is usually not prepared). As much as happy surprises are welcomed, just as little welcome are unhappy surprises that plunge the planning process into confusion and create chaos. But, it is quite unreasonable to think that these or any comprehensive and difficult task would be free of surprises.

## 5.6 Robustness of Decisions

Strategies can be understood as guidelines for future proactive activities and the related decisions. Once established in a common vote, the guidelines should help the actors not to lose sight of the goal and actively work on realising it, because most of these processes take many years to implement with the desired development orientation.

Because, as a rule, strategy changes are very complex and work-intensive, once a development orientation has been chosen, even with changed spatial conditions, if possible, it should be maintained. Gradual realisable solutions with coherent stages promote mobility and help limit the risks connected with changes of the spatial conditions. Such solutions are robust and they should have preference before any all-or-nothing solutions that react very sensitively to a change of circumstances and framework conditions. Robust solutions allow an increase in the indefinite elements of information. In other words: *Changes in the essential framework conditions do not inevitably lead to a change in the chosen path of the solution.*

Useful methods should be applied to increase the robustness of decisions and consequently remind us that knowledge can be incomplete, especially for decisions that reach far into the future, surely a legitimate reminder Scholl [7]. One can approach them pragmatically because step-wise realisable coherent, upwardly compatible solution building blocks are being sought. Taken together, these building blocks should give a coherent whole, but each should also give a glimpse of a realisable and valuable contribution into the problem to be solved. Neither should the individual building blocks of future steps prejudice or hinder the process. Early deliberation on such connections and complicated pieces of the puzzle often open up new possibilities for combinations and variations Popper [4]. The more possibilities exist, the more flexible the reactions to changes in the spatial conditions. In other words: *The better the planning, the more open the future can be.*

## 5.7 Integrated Development of the European North-South Railway Link

Using the North-South Railway Link (Rotterdam–Genoa) as an example, I would like to make the methodological considerations clear. Part of the transit space is the Rhine River axis from the North Sea to Basel with its bordering regions of the Netherlands, Germany and Switzerland.

Over the entire history of Europe and its spatial development, this axis has formed a quasi backbone. The most densely settled areas of Europe are to be found here, together with diverse landscapes and urban regions of various sizes. It is also the area of the largest economic value creation. To a considerable extent, the economic power of this axis determines many European developments. Major investments of all kinds: airports, railway systems, distribution centres, industries, etc., take place here.

The inner renovation and development of the regions in this corridor are a challenge for those involved. Global, European, national, regional, and local processes overlap each other. The changes are plain to see. The serious conflicts with spatial development in this transit area also lead inevitably to addressing Europe's cultural and political history as well as its future and the future of its nations. What are the central questions, and what are possible solutions?

In a small informal circle in Switzerland, we call it the *Spinnerclub*,[2] we began to discuss interrelated questions about border issues. After some time, we established that the problem definitions in the individual parts of the Corridor were similar: unsolved problems of noise protection, massive setbacks, e.g. because of financial limitations in the commissioning schedule and, against this background, possible

[2] In English, this might considered an informal think tank.

capacity bottlenecks if goods transport should increase sharply. This in turn might mean that the development of regional transport on the railways might be checked, blocked, or even totally eliminated, which would contradict the settlement policy of redevelopment. This assumes from the beginning that stations and stops of public transport would be crystallization points for this development. Questions upon questions. What followed was a decision to conduct a cross-border test planning process on spatial and railway development in combination with the Regional Association of the Southern Upper Rhine, who had joined us for the first time.

This act of cooperation contributed to the initiation of railway and spatial development in Corridor 24 (Rotterdam–Genoa), which later became a European Union project and is now designated as a strategic project by the EU. But what does 'strategic' mean in this context?

## 5.8 On Spatial Strategies

Strategies are guidelines for the implementation of an integrated spatial solution that [10, 11], once approved, may take many years or even decades to realise. An overview and focus points are necessary elements of a nascent strategy, while problems, understood here as difficult unsolved tasks, are the central starting point (see Schönwandt [13]).

Tactics are concerned with the details of an operation, while strategies bring more flexibility into constantly changing situations and take existing uncertainties under consideration. The approach to limited means and resources, along with risks and surprises, plays a central role. Strategies always contain deliberations on the use of restricted resources: time, financing and qualified specialists. The use of strategies recognises that even with the best planning, reality can never be completely controlled. Thus, the availability of reserves is a central precondition for being able to solve difficult tasks.

In order to be able to develop strategies at all, clear periodic evaluations of the situation are needed. These call for the creation and evaluation of important overviews, a clear appreciation of the central tasks, examination of the results and, finally, concentration on the essentials needed for the next period. Site evaluations lead to decisions on the use of resources and limited resources force a concentration on the essentials. A strategy helps to differentiate the essential from the unessential. Strategic thinking is not a new planning theory; it is an expansion of the approaches already available.

## 5.9 The European North-South Railway Link: Main Artery of the Railway Network

The European North-South Railway Link from Rotterdam to Genoa is one of the most important railway connections in Europe. The 1,200 km transport route has grown up over centuries and today connects the largest North Sea ports of the

Netherlands and Belgium with Germany, Switzerland, northern Italy and the Mediterranean. The main goal of the railway link is to attract cross-border railway transport and to transfer trans-alpine goods traffic from roads to rails. The core of this system is the level railway links through the Alpine Base Tunnels, which will increase train speed and volume.

Major investments in railway-related infrastructure of more than 40 billion euros will be undertaken along this axis. In Switzerland alone, the construction of two new tunnel routes will require investments of about 15 billion euros (including noise protection).

The opening of the first Alpine Base Tunnel, Lötschberg, in 2007 and the planned opening of the Gotthard Base Tunnel at the end of the 2016 together with Ceneri Base Tunnel at the end of 2019 constitutes a quantum leap in railway development, and not only in the alpine area. Tunnel construction, including its approach routes, should bring with it favourable connections for cross-border goods and passenger transport in the central European North-South Corridor (www.alptransit.ch).

Two additional tunnels in the Alpine EU area, namely, the Brenner Base Tunnel and the Mont Cenis Base Tunnel in France, are also planned as a component of the Trans-European Transport Network (TEN). The Mont Cenis Tunnel between Lyon and Turin is still in the planning stages and the construction of the approximately 50 km Brenner Base Tunnel started in early 2011, according to the plans of the Brenner Construction Consortium (Figs. 5.1 and 5.2) (http://www.bbt-se.com).

The North-South Link runs through European regions that have a high added-value rate, a strong transport tax revenue on major rail and road networks and a high settlement density. Nearly 70 million people, approximately one-fifth of the expanded EU of 2004, live within its catchment area and annually 700 million tons of freight move from the northern ports to its hinterlands and south through the Alps, which is over 50 % of the total freight travelling north-south across Europe (Fig. 5.3).

### 5.9.1 Considerations on the Alpine Area

The conclusion can be drawn that the base tunnels of the Gotthard and Ceneri Passes will be in operation more than a decade earlier that the Brenner Base Tunnel. It can be assumed that logistics and cross-trade efforts will be employed earlier on this site and will be correspondingly positioned. Whether and which relocation will happen after the commissioning of the Brenner Base Tunnel, cannot be predicted at this time. However, it can be reckoned that well-rehearsed operational procedures, investments already made and the good geo-strategic position of the NEAT to the economic centres in northern Italy as the shortest connection to the northern harbours will prove to have a certain inertia. Basically, from a European viewpoint, it will not be a disadvantage when redundant base tunnel systems are available for this important north-south relationship. The base tunnels are a long-term investment in the future of Europe.

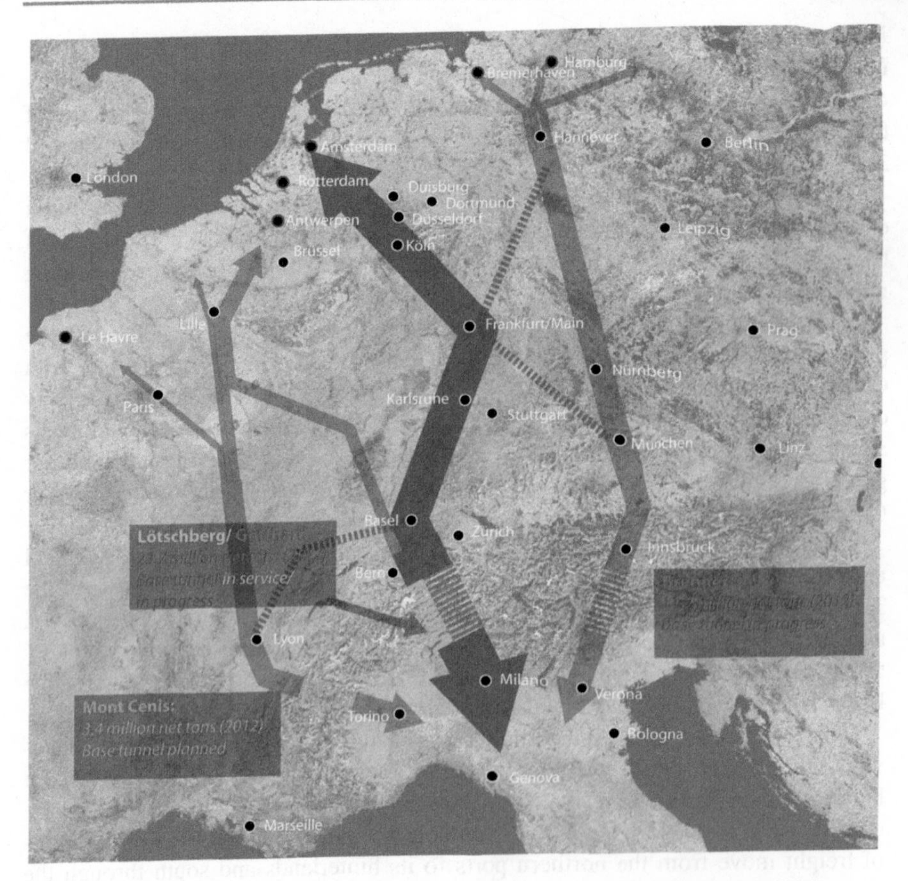


**Fig. 5.1** Planned and in-progress Alpine Base Tunnels (*Source* IRL, ETH Zurich 2006)

## 5.9.2 Importance of the Harbours

A strong increase in the transport from north to south is expected in 2030. This reflects the continuing globalisation movement. The world grows together slowly, but surely, to one single large market. Buying, production, selling or marketing will take place where it can achieve the highest advantage. Economic growth in Asia, mainly in China, will play an especially important role.

The main receivers of ocean-going transport are the ZARA harbours (Zeebrüge, Antwerp, Rotterdam, Amsterdam) and the German harbours (Wilhelmshafen, Hamburg, Bremen/Bremerhaven). In 2010, Antwerp and Rotterdam handled over 16 million TEUs (twenty-foot equivalent unit) of cargo. By 2020, capacity is expected to double to over 36 million TEUs.

The question is often asked why there are not more ships using the southern harbours, in particular, the Ligurian ports of northwest Italy, instead of shipping all the way to the northern harbours. The preference for the northern ports can be

**Fig. 5.2** Breakthrough in the Gotthard Tunnel 15 October 2010 (*Source* Alptransit)

traced back to the much larger and more effective cargo facilities and better inland connections: water routes, railways and highways. For example, just one number: the container capacity of Rotterdam alone is five times larger than all the Ligurian harbours together: Savona, Genoa and La Spezia. An additional consideration for the future is that some of the North Sea harbours, in contrast to the southern harbours, are preparing to serve as deepwater ports for container ships with a large cargo handling capacity (ca. 18 thousand TEU per ship), currently only Rotterdam and, in preparation, Wilhelmshaven. At the present, one reckons with ships from 12 thousand TEU, which corresponds to a load of about 120 trains or 6,000 lorries.

The harbours of northern Europe, mainly in Rotterdam, are investing heavily in the development of their harbour infrastructure. Rotterdam will increase its cargo handling capacity from almost 450 million tonnes at present to 700 million tonnes by 2035 with the construction of Maasvlakte 2. The harbour extension will then add up to ca. 50 km. However, the harbour construction will make it increasingly more difficult to guarantee a congestion-free inland transport. Already traffic jams are a daily occurrence in the areas surrounding the harbours and its target destinations.

One demand on the Rotterdam harbour consists of moving part of the container inland transport onto water routes (Rhine River), starting in 2009 to increase from 40 % (1.6 million TEU) to 49 % (8.2 million TEU) and increase railway transport from 14 % (0.5 million TEU) to 20 % (8.2 million TEU). The lorry share should be reduced from 46 % (1.9 million TEU) to 35 % (6.4 million TEU). Measured by the construction goals of the harbour, a ship will carry five times more TEU, lorries four times more and trains sixteen times more.

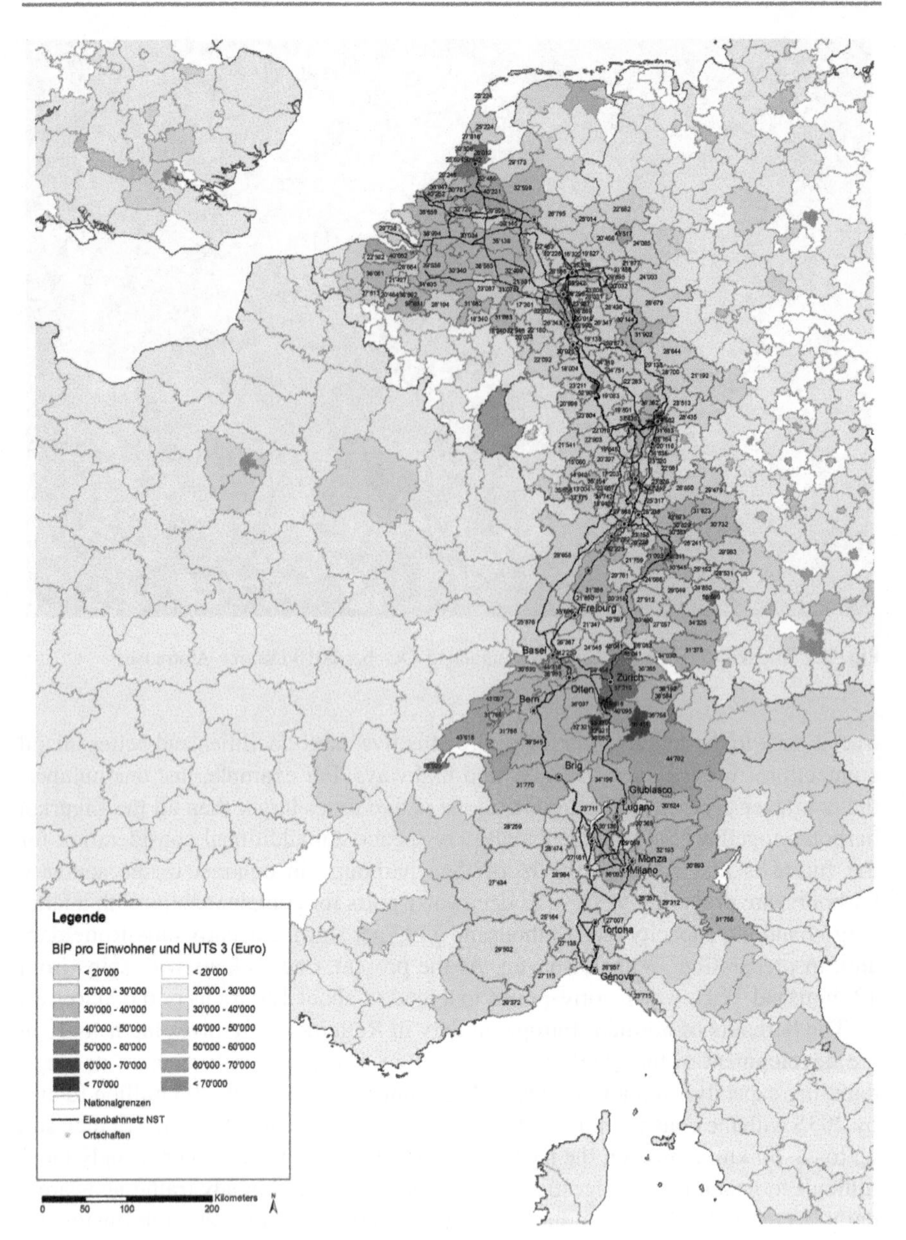


**Fig. 5.3** European regions with the highest added value (*Source* IRL, ETH Zurich)

A further increase in the demand for transport over the North-South Link is expected by 2030 (Fig. 5.4). However, since the ports in northern Europe, mainly Rotterdam, are investing massively in the construction of infrastructure, it will be increasingly difficult to transfer traffic to the hinterlands.

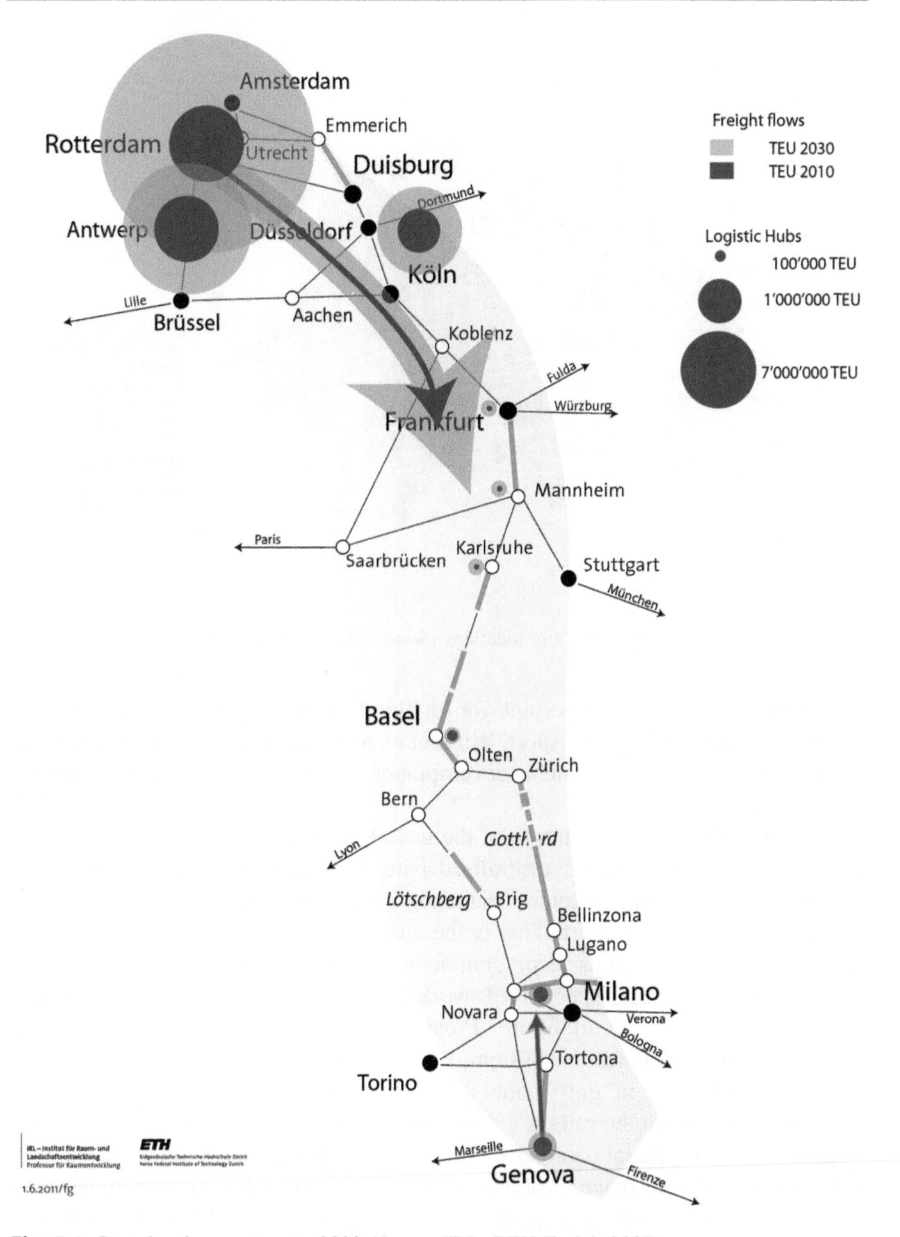


**Fig. 5.4** Port development up to 2030 (*Source* IRL, ETH Zurich 2007)

As mentioned, the investment for the railway-related infrastructure of the entire North-South Link requires major resources, approximately 40 billion euros. These resources are not available in the individual countries or in the EU budget. It is becoming apparent that countless bottlenecks will continue to exist because not enough capacity will be available for all the desired rail transport. There is a risk

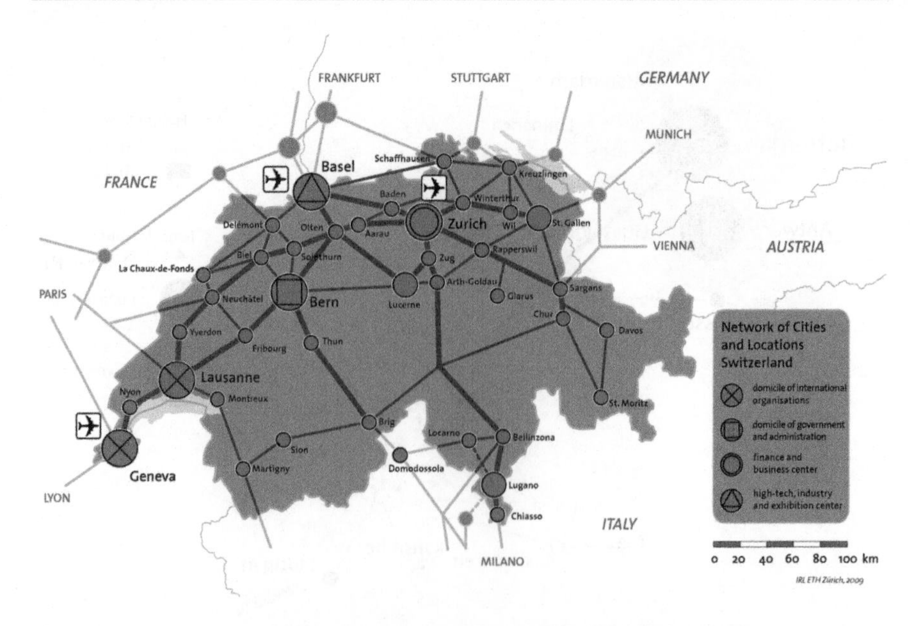


**Fig. 5.5** Swiss network of cities and locations (*Source* IRL, ETH Zurich 2007)

that regional transport, important for public transportation, will be displaced through increasing goods transport. Without an attractive public passenger service, the desired goal of settlement development through redevelopment will be endangered.

Switzerland is a good example of the use of a strategy in its network of cities and places (Fig. 5.5). The decentralised pattern of settlement in the country, a reflection of its federal structure, demands an efficient public transport system, mainly through rail transport. This is the strategic backbone of its spatial development, because through its secure, efficient and well-operated system, mobility is ensured for the highly specialised workplaces in Switzerland and bordering countries both now and in future. Prosperity and competitiveness are closely related to this transportation system, even when its contribution to the total transport performance is 'only' about 20 % of total passenger kilometres travelled. Scope for mobility on the rails is a contribution to the social stability of the nation [8] and its historic spatial paradigm. Thus, social networks can be maintained even with a change of workplace. Problems of excessive concentration of population in less densely settled centres could be avoided, as with the problem of the depopulation of peripheral areas.

Naturally, cities and other locations of the Gotthard axis also belong to this network. From Basel to Chiasso, approximately 3.6 million residents live in its catchment area. The majority of the Swiss population lives in Central Switzerland, which has an east-west direction. Improvements in capacity and speed would mainly improve passenger transport. However, the major part of the goods transport will be in the north-south direction. Today, the Gotthard axis (Basel–

Chiasso) already takes more than 70 % of rail goods traffic (transit 55 %, import-export), which predominantly travels over the Basel–Chiasso route. With the opening of the Gotthard Base Tunnel, this number will rise to about 80 % and transported goods will rise from 25 million tonnes to 35 million. At the same time, important destinations in the north (city networks on the Upper Rhine, Frankfurt, Rhine-Main and Rhine-Ruhr, Randstad Holland) as well as south (mainly the Milan area) will be connected over the north-south axis.

The further development of the railway network is not about the construction of new lines, but more about sufficient and reliable capacities of the main routes. For this reason, Switzerland has invested heavily in this infrastructure. At present, this is about two billion CHF per year, almost as much as Germany, which is ten times larger, has invested in the total railway sector. Switzerland is also facing diminishing resources, which is why clear priorities must be set. Securing the quality of the stock and its robustness as well as a measured increase in service capability must therefore have priority over an increase in speed, i.e. building new routes. In spring 2014, Swiss voters will have the opportunity to vote on a financing fund for public transport. This will create, if the guidelines are accepted, mid- and long-term planning security for the financing.

Switzerland should keep the financial situation in Europe in mind, however, because the lack of financing could mean delays in both the northern and southern access routes to the base tunnels—even with contractual agreements. For the northern access roads, this means stronger pressure on the Rhine River for goods transport. According to experts, there is still a reserve of up to 20 % available. For this reason, the Rhine ports of Basel-Muhlhouse-Weil (cargo handling in 2010 ca. 12.5 tonnes, 185,000 TEU), in particular, the Swiss ports of Muttenz, Birsfelden and Kleinhünigen, will become more important as gateways for goods transport to the north. For the last port, Kleinhünigen, there is the possibility to rebuild it into a single tripodal harbour, from which someday goods could be transferred directly from the water to a long-distance freight train. This would allow goods to travel from the North Sea via the Rhine and a trimodal harbour, if necessary, and thus 'sail around' the railway and roadway bottlenecks of the north-south axis.

The option of a loading terminal for trucks to transport goods over the alpine crossing should be kept under consideration by the planners, even though the Swiss Constitution states that goods must be placed on trains. Another reason to keep an eye on the truck terminal is climate change in relationship to the frequency and intensity; it may lead to changes in high and low water levels and thus reductions in shipping.

At this time, one assumes that in the summer more low water level situations would appear. In the southern process, it must be considered for a major part of the arriving goods trains. It must also be considered that a major part of the goods on trains in southern operations must be reloaded onto lorries. The transfer systems in the southern border areas will therefore gain special importance. In case the southern part of the Gotthard route is ready very much later than the commissioning of the Gotthard Line (thus after 2019), a massive overloading of the high-capacity roads in the Greater Milan Area must be reckoned with.

### 5.9.3 Strategies for Integrated Spatial Development

The example of the North-South Railway Link clearly illustrates that there are many questions still unresolved about the realisation of this strategic trans-European connection. If this railway connection cannot be realised and sufficient capacity does not become available, there is the danger that the quality of development in some of the most intensive wealth-creating areas of Europe will deteriorate significantly. The major in-process investments for harbours and logistics structures in northern Europe are being carried out detached from a clear strategy for the Corridor (Fig. 5.6).

The following elements, which include the results of a CODE 24 conference at the ETH Zurich (ETH Zurich [14]), could lead to a collaborative strategy for the North-South Link (Corridor 24):

- Europe must invest heavily, mainly in rail infrastructure and in multi-modal transfer facilities if the continent wants to remain globally competitive.
- All efforts must be oriented towards this goal in order to be able to take steps at the right time to prevent the expected bottlenecks in various countries.
- Goods transport must be transacted in an efficient and environmentally friendly manner. As much as possible, it should be transferred to the rail system.
- For an economical approach to the use of land and an efficient approach to mobility, settlement development needs to be oriented towards the catchment areas of the public transport system infrastructure.
- For the functionality of the railway infrastructure over the entire corridor, it is important that the Region Rhine-Ruhr can fulfil its role as a trimodal logistic hub for the transfer of goods from the river to the rails in increasing volumes.
- The Middle Rhine presents strongly conflicting interests between the local population and the global goods stream, the noise situation especially is leading to major conflicts. New solutions must be found if the transfer goal is not to be endangered by a lack of acceptance of the population affected.
- A mutual view of the Frankfurt-Mannheim area, the most over-burdened section of the Corridor, is important for its further development. This region is central, not only for the flow of goods, but also for the logistics areas for cargo handling between the North Sea harbours and the most important industrial centres. Essential facts for the development of the Corridor are, however, still not clarified and need a deeper integrated examination.
- The southern part of the Upper Rhine offers the most heavily used two-track route with its Rastatt section, however, a solution for this bottleneck is conceivable. Essential questions remain, such as how the embedding of the infrastructure in the cultural landscape in Offenburg and further south can take place in order to meet the growing resistance of the population.
- The large capacity and productivity gain that will follow the new Alpine Base Tunnels, will shrink because of the partly deficient capacity on the feeder routes, as well as the planned intensive use of the route for passenger transport.

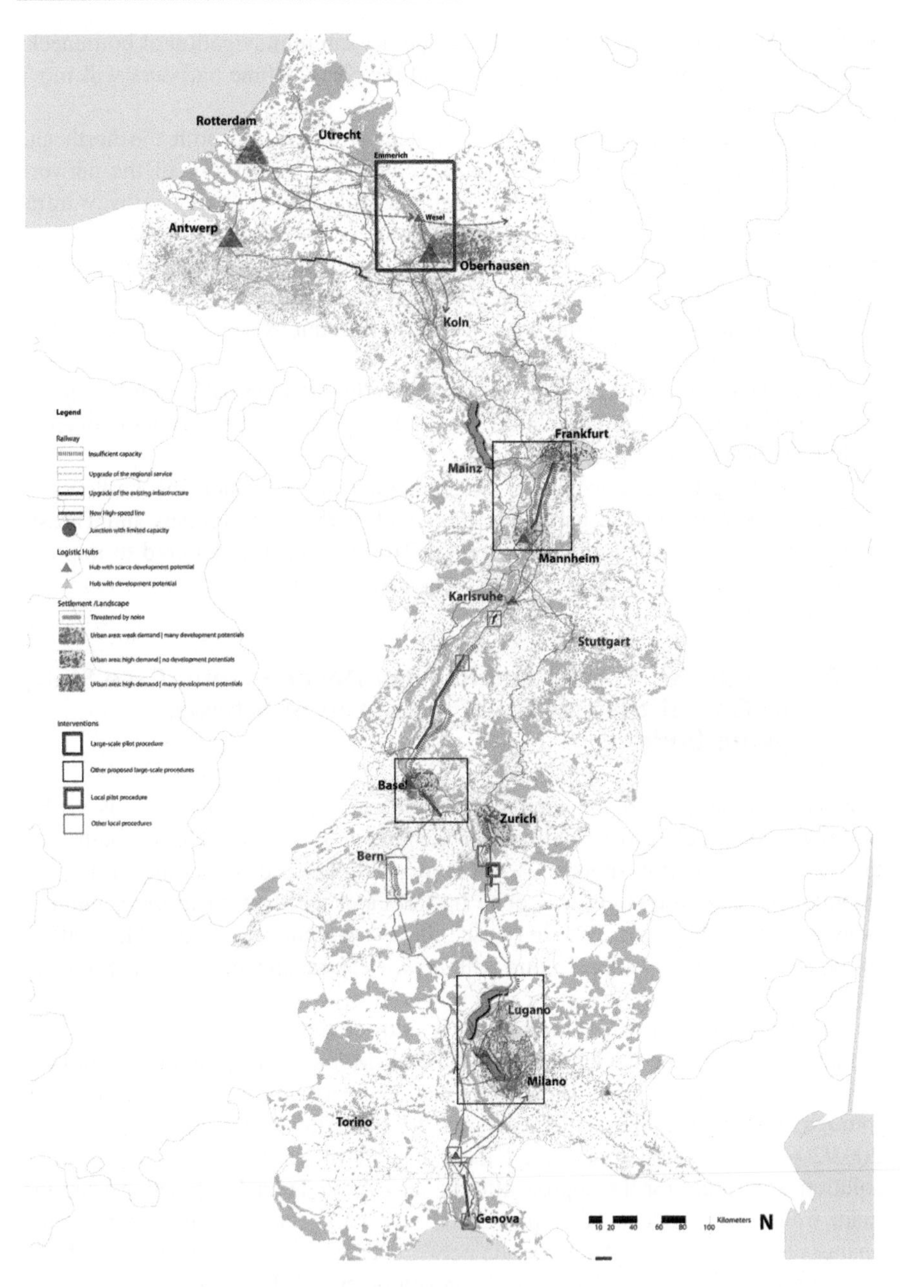


**Fig. 5.6** Overview map of the site evaluation (*Source* ETH Zurich IRL for INTERREG IVB NWE project CODE24)

- Using the Rhine water route for the temporary circumnavigation of bottlenecks in Germany will grow in importance and the Swiss Rhine harbours will move into focus.
- A lack of coordination of the development strategies on both the north and south borders of Switzerland will also reduce the capabilities of the network nodes and their access paths. A coordinated plan for planning railway infrastructures and spatial development is absolutely essential.
- Missing infrastructure is not a major problem for hinterland rail traffic to and from Genoa; the city is connected by five track lines. The essential limiting factors are to be found in the processes in the harbour as well as in the operation and technology of the railway.
- The collaboration between the various Italian harbours remains to be explored. Here, a strong collaboration with Livorno, La Spezia and Savona could open up interesting perspectives.
- For the long-term, it would be of great interest to include the harbours of southeastern Europe, in particular, the Greek harbours of Pireaus and Thessaloniki, in the overall European strategy. To this end, it is required to improve the railway axis between Pireaus/Athens and Vienna.

## 5.10 Integrated Spatial and Railway Development: Case Study of One of the Hot Spots in Canton of Schywz: Felderboden

Hot spots are designated areas in the Corridor where problems are already visible and where comprehensive solutions for the functioning of the North-South Link are needed. Such hot spots are spaces of national and sometimes European importance. The case study was done on the Felderboden area in the Canton of Schwyz, a local hot spot. The study used the test planning method, which will be introduced later in this article along with how it can be applied to resolve difficult problems and tasks.

The development of the Urmiberg axis is of strategic importance for the settlement and economic development of the Rigi-Mythen region and the Canton of Schwyz as a whole. Infrastructures of national and European importance run through this main alpine settlement area. In an earlier test planning process in 2002/2003, the canton reorganised the transport infrastructure. In particular, a solution was found for the approach to the new Axen Tunnel, a junction from the main trunk line that can be implemented gradually, and combined with future changes to the railway and automobile network to save space. These are now basic conditions for any further ideas for the development of this area.

Among the development areas of the Urmiberg axis is a large contiguous plot of land with above-average individual plot sizes. In the second test planning process in 2011, the canton provided a strategy for an integrated spatial and infrastructural development in a long-term perspective up to 2030. The development of this area can be pursued within the spatial planning principle of redevelopment before new development because the area lies close to the centre and already has a relatively favourable position to the local public infrastructure system. An area of ca. 25 hectares is available within the perimeter of the contiguous premises that would allow the step-by-step realisation of ca. 100,000–150,000 $m^2$ gross floor area in the Brunnen Nord area alone. An important consideration is that in the surroundings of the Urmiberg axis, and dependent on it, investment in the order of more than 2.5 billion euros (at June 2011) will be transacted in the next ten to twenty years. Parts of this will be the realisation of the Morschacher Tunnel, the renovation of the existing highway network and the gradual implementation of the NEAT (New Railway Alpine Transport).

The area lies on the main artery of the north-south transport route, which provides easy access and will give impetus to the marketing of the area. Good transport connections are an indispensable prerequisite and help develop liveable alpine communities in the area. The realisation of the NEAT approach routes in the Canton of Schwyz with a connecting junction in Felderboden opens the way to relieving the SBB main lines of goods transport. The test plan results recommended pursuing this plan in order to achieve the earliest possible realisation. This must be considered in the context that a high-capacity connection in the direct approach to the Gotthard Base Tunnel is of extraordinary importance for a robust and secure operation of the entire north-south rail connection. It should be noted that work begins at the earliest in 2016, but may only start after 2030.

The Steering Committee recommended concentrating future settlement development around the two poles of the Urmiberg axis, Seewen in the north and Brunnen in the south, to create a green space in the middle, free from further construction and outside of the already zoned and built areas. Keeping landscape space away from construction also puts the settlement poles in a better marketing position. The basic thinking behind the concentration of future settlement development corresponds to the general principle of 'redevelopment before new development'.

## 5.11 Learning by Doing: The Test Planning Method

The case study demonstrates that integrated solutions for difficult tasks can be developed within a manageable time frame. This method, tested on many different kinds of difficult tasks, can lead to integrated solutions and breakthroughs. The core of the test planning process is based on certain principles and enables the pros and cons of solutions to be discussed directly and immediately, or as we say: simultaneously. The testing of such ideas in the interplay of suggestions and critique allows basic solution ideas to crystallise within a time period of about nine months.

Space does not allow a detailed description of the entire test planning process. Full descriptions can be found in numerous other publications [9–12]. This method fits in with the democratic structure of the European nations.

## 5.12 Democracy and Spatial and Transport Development

As experts, democracy forces us to find the most possible impartial alternative solution for difficult tasks. We manage this best, in my view, when we are ready at the very beginning to allow ourselves to engage in the rivalry of competing ideas and to illuminate the entire bandwidth of possible solutions, in short, to test it.

The point of the process is to identify the arguments for and against a particular solution as clearly as possible. The best way to manage this, according to the philosopher Popper [5, p. 273] is to invent courageous hypotheses to solve difficult problems and expose them before an entire arsenal of critics. In the Pro and Con arguments, which are best delivered first in the direct discourse of point and counterpoint by the experts, the solutions that are advisable will crystallise out, as well as those that are not at all suitable. This aspect is also important in that the suggestions that are not worth following can be put aside with good arguments leaving those worth following to be recommended to the political authorities. It has been said that the experts should not overestimate themselves; they are not obligated to take the material decisions. However, they should also not underestimate their work either. In hindsight, the most far-reaching mistakes are usually made right at the beginning—and can come from any quarter.

All this work needs time. Because of this, many think that democratic processes are slow and nowadays slowness is seen as a problem. But the larger factor of having legitimatisation is a value that usually proves to be a higher value than the speed of the decision. A slower process often provides protection from decisions and errors made in haste. And in planning, with its many different levels and players, a slower pace can be the faster way to put a plan into practice.

## References

1. de Jouvenel B (1967) Die Kunst der Vorausschau. Berlin, Luchterhand, p. 22
2. Hesse M, Beckmann K, Heinrichs B, Scholl B, et al (2010) Neue Rahmenbedingungen, Herausforderungen und Strategien für die grossräumige Verkehrsabwicklung. In: Forschungs- und Sitzungsbericht der ARL. Akademie für Raumforschung und Landesplanung, Hannover
3. Modigliani F, Cohen KJ (1961) The role of anticipation and plans in economic behavior and their use in economic analysis and forecasting. Urbana-Champaign, Illinois
4. Popper K (2000) Vermutungen und Widerlegungen: Das Wachstum der wissenschaftlichen Erkenntnis. Mohr Siebeck. Auflage: Einbändige Studienausgabe der Bände 1 & 2
5. Popper K (1984) Objektive Erkenntnis. Ein evolutionärer Entwurf, 4th edn. German translation by Hermann Vetter. Hamburg. Original title: Objective knowledge, Oxford 1972
6. Popper K (1976) Ausgangspunkte. Meine intellektuelle Entwicklung. Hofmann und Campe Verlag, Hamburg
7. Scholl B (1995) Aktionsplanung. Zur Behandlung komplexer Schwerpunktaufgaben in der Raumplanung. Vdf Verlag, ETH Zürich, Zürich
8. Scholl B (ed) (2007) Perspectives on spatial planning and development in Switzerland. Report of the international group of experts. Bern: Swiss Federal Office of Spatial Planning (ARE). Also available in German, French and Italian

9. Scholl B (2010) Testplanungen als neue Methode. In: TEC21. Fachzeitschrift für Architektur, Ingenieurwesen und Umwelt, Nr. 29–30
10. Scholl B (2011a) The relationship between sector planning and spatial planning. In: SAPONI, spaces and projects of national importance. VDF Verlag, ETH Zürich, Zürich
11. Scholl B (2011b) Strategies for integrated spatial development along the European North-South Railway Link. ISOCARP Review 07, The Hague
12. Scholl B, Staub B (2013) Test planning—a method with a future. Cantonal Office of Spatial Planning, Solothurn
13. Schönwandt W, Grunau P (2013) Planen und Entwerfen als Lösen komplexer Probleme. University of Stuttgart, Stuttgart
14. Zurich ETH (2013) Professorship for spatial development. Bericht Workpackage, Zurich 2013

# Evaluation Tools to Support Decision-Making Process Related to European Corridors

6

Isabella M. Lami

**Abstract**

In many European Countries any decision to draft a plan, to define a path of transport infrastructure or to choose the location of an "undesirable service" requires an imposing volume of discussions and negotiations. The most worrying aspect is that, even if the processes of governance are sufficiently open and transparent, they can run the serious risk of failure, as shown by recent experiences. In an attempt to reduce this risk, the decision process has to be seen as the result of a set of interactive actions occurring at different times in order to distinguish what is a priority and what may be negligible. The view of planning as a strategic choice process is a dynamic one, which implies to choose in a strategic manner rather than at strategic level. The chapter, after framing the main evaluation tools adopted in the field of territorial transformations connected to mayor transport infrastructure (as the Eurocorridor), provides some insights about the choice of the most suitable MCDA methodology. It introduces also the idea of MCDA in combination with visualisation tools to tackle these types of decision problems. Many and varied experiences of using multicriteria as tools to support decision aiding processes in a European project are illustrated in the last paragraph. It is shown how they have stimulated general reflections with the possibility of increasing the affordances, i.e. the possibilities for action the methods offer to those involved, varying the degree that was crucial to enable or constrain (model-supported) meaning negotiations and new knowledge creation.

I. M. Lami (✉)
Department of Regional and Urban Studies and Planning (DIST), Politecnico di Torino, Corso Massimo D'Azeglio 42, 10123, Turin, Italy
e-mail: isabella.lami@polito.it

I. M. Lami (ed.), *Analytical Decision-Making Methods for Evaluating Sustainable Transport in European Corridors*, Sxi 11, DOI: 10.1007/978-3-319-04786-7_6,

## 6.1 Introduction

In many European Countries any decision to draft a plan, to define a path of transport infrastructure or to choose the location of an "undesirable service" requires an imposing volume of discussions, negotiations and arrangements. At the same time, these may be characterized by the protest of local communities fighting for the right to choose what happens to the land they inhabit [25, 45], to the point that nowadays territorial conflicts have become more frequent, widespread and often even more disruptive than social conflicts.

As Bobbio [4] underlines, as soon as a problem arises, the first reaction from the local public administration is to open a discussion: public decisions are the result of a continuous negotiation process which concludes in reaching agreements. Escalating transaction costs represent the main obstacle to the territory's government. In a situation of high institutional and social fragmentation, the power of veto is in fact multiplied. It does not refer only to the traditionally strong interests, but also to the traditionally weak interests (as long as there is a concentration). Groups that are not involved in the decision process have the possibility to stop the choices made by others, or at least to delay them.

The most worrying aspect is that, even if the processes of governance are sufficiently open and transparent, they can run the serious risk of failure, as shown by recent experiences of territorial transformations. For these reasons, is it possible to affirm that "decision is not an act but a process" [48] characterized by continuous learning. The decision process has to be seen as the result of a set of interactive actions occurring at different times in order to distinguish what is a priority and what may be negligible. The view of planning as a strategic choice process is a dynamic one, which implies to choose in a strategic manner rather than at strategic level. The concept of strategic choice is related to the connectedness of one decision to another in a continuing dilemma of balancing urgency against uncertainty in decision-making over time [18]. It is in this theatre of complex interactions that it has been generally agreed that Multiple Criteria Decision Analyses (MCDA) can provide a very useful support. Belton and Stewart [3] define multi-criteria decision analysis (MCDA) as ''an umbrella term to describe a collection of formal approaches which seek to take explicit account of multiple criteria in helping individuals or groups explore decisions that matter''. Assuming that the MCDA could really contribute to the decision processes which refer to the transport infrastructure, the chapter provides some first indications. It aims to suggest "when" and "how" MCDA could be applied. Significant insights are shown throughout the direct experiences in the CODE24 project, and it will be shown that these experiences are generalizable.

The chapters are structured as follows. After the introduction Chap. 2 highlights the current phenomena of the territorial conflicts with particular reference to the transport infrastructure. Chapter 3 frames the main evaluation tools adopted in this specific field and provides some insights about the choice of the most suitable MCDA methodology (the illustration of a comprehensive list of all the MCDA is therefore beyond the scope of this chapter; though some specific references are given). Chapter 4 analyzes the experiences of decision support referring to territorial and transport problems along the corridor Genoa-Rotterdam, developed in the Code24 framework, where some general reflections are provided. The conclusion shows possible development in the field research.

## 6.2 Territorial Conflicts Concerning Big Transport Infrastructures

In recent years, several trends affecting the nature of the policy-making process can be observed [44]:

1. An increasing demand for participation coming from the citizens;
2. An increasing mistrust between citizens, policy makers and "experts";
3. A growing social fragmentation;
4. A rising scepticism towards science;
5. A rapid growth in the amount of information available often provided without assessing reliability.

The result of such trends is an increase in the demand for accountability and legitimation, for both the process and its outcomes.

The decision-making processes in the field of territorial transformations are forming into specific characteristics [11]. The first of such characteristics is the increase in complexity, with an expansion of decisional network both, at geographical scales and at the level of relationship between public and private actors. In this sense, new types of actors come into the decision-making arena, alongside traditional ones. The result is a pluralisation of points of view within the processes, with a progressive separation between the ways in which public decisions are taken and what is required by the constitutional laws. The second distinctive characteristic concerns the increase in uncertainty and, in particular, the uncertainty about the outcomes of the decisions [18, 34]. Finally, there is an increase in conflicts among social groups, political actors, citizens and public authorities.

Since the territorial conflicts have become so important, it is significant to analyse why they have increased in the last decades and what is at stake. Bobbio [5] gives six types of interpretations to these questions. The territorial conflicts are seen, from time to time: (a) as the expression of particularistic and egoistic points of view that prevent the fulfilment of the general interest; (b) as the pressure of vested interests that exploit the fear of the population for other purposes; (c) as the consequence of the imbalance between concentrated costs and distributed benefits; (d) as a reaction to risks that are deemed unacceptable; (e) as the resistance of the places against the flows that invade or cross them; (f) as a demand for a different model of development.

In the specific context of the location and construction of big transport infrastructures, like the Eurocorridor, particular interest and focus has been given to points (c), (d) and (e); as described and illustrated by Bobbio [5].

The conflict can be seen as the result of an effective and predictable physiological imbalance between the costs and benefits of the proposed settlement. If the benefits are widespread and the costs are concentrated on a small community that is forced to bear the costs of an intervention that benefits others, it is quite natural that conflict arises. The approach to resolve the conflict lies in the negotiation: the proponents and communities involved negotiate the mitigation and compensation necessary to make the project acceptable. This strategy has the advantage of taking the arguments of opponents seriously, but is met with some difficulties. The promoters are often willing to offer compensation, but they are less willing to revise their plans and this attitude creates the unpleasant suspicion that they want to "buy" the health and safety of said community. In addition, local communities are composite entities and it is not easy to find the right person with whom to negotiate.

Another interpretation is that territorial conflicts are the direct consequence of new fears stimulated by technological development. The object of the dispute is, according to this narrative, the nature of the risks associated with a settlement, while the solution of the conflict would be the elimination of these risks or, at least, the definition of which risks are of minor importance or unlikely and therefore become acceptable.

If governments do not consider it necessary to inform the public and involve them in the political procedure, citizens can now appeal against these decisions in order to feel part of the process. In the risk assessment there are important psychological aspects that come into play: the feeling of being consulted, participation in decision processes and the impression of having the power to change things [32]. This may contribute to the understanding of individuals' perception of risk. This contention is difficult to solve. Ordinary citizens have a perception of risk differing greatly to that of specialists. They most fear the risks that are imposed on them rather than those they have voluntarily chosen themselves. Risks that are poorly understood, which are highly unlikely but catastrophic, are also reasons to cause fear. However the risks that are less damaging but more likely are of little concern. They also have risks in mind that specialists tend not to recognize: the depreciation of property, the consequences for the local economy and quality of life. Supporters of the interventions try to show—with standard arguments based on the calculation of probabilities—that the actual risk is different from what is feared and accuse their opponents of cultivating unscientific and irrational fears. But the specialists are unlikely to breach the concerns of the counterparty because such reassuring predictions have often proved unfounded in the past and because the risks feared by those who protest are different in nature than those of the specialists. It should be noted that the fears—even if unfounded—generate very concrete consequences, i.e. panic flooding the stock market or, for Corridor 24, the fall in real estate values in risk-prone areas, if the noise pollution of high-speed trains is fully mitigated, there is still a concern that an increase in train frequency or a development in technology can alter this situation making an area unattractive to purchase a home in the surrounding area.

As for the last interpretation, in the specific context of the construction of a transport infrastructure, the territorial conflicts can be seen as the resistance of the territories against the flows that invade or cross them [5, 46]. Globalization has made every border permeable, multiplying the flow of people and goods and increasing the susceptibility of those who are exposed to the currents of these crossings. The conflict between flows (in constant motion) and places (static) is one of the dominant traits of our time [10]. Not all flows are unwelcome. The regions/cities are competing to attract beneficial flows (investment, universities, tourists, etc.); and at the same time, try to drive away unpleasant flows (poor foreigners, waste treatment plants, power plants, etc.). The territorial conflicts are the manifestation of this competition. Beyond the actual dangers that the flows are likely to generate, receiving an unpleasant flow could lead to a derating of local territories. The object of the dispute, according to this interpretation, is sovereignty: global versus local sovereignty. The territorial protests, when they manage to hold up over time, become identity movements. The identities appear as non-negotiable values [5].

To summarise, one can say that in a transport planning context unstructured problems need to be addressed. These problems are characterised by the existence of [35]: (1) multiple actors; (2) multiple perspectives; (3) incommensurable and/or conflicting interests; (4) key uncertainties.

For all of these reasons, the need for decision support tools that are able to consider different aspects of transport planning is becoming increasingly more evident, overcoming the logic of simply applying the cost-benefit analysis, which has been until recently, almost the unique assessment tool in the field of transport [29].

## 6.3 Evaluation Methods for Transport Policies and Projects

A territorial transformation could be seen as a search to balance needs, institutional and financial constrains and market responses, within a perspective of sustainability.

With this idea, evaluation tools seem to be essential in order to control the complexity of the system and to support the governance of the transformation. During recent years, the evaluation approaches tried to consider the progressive complexity of urban and territorial transformations moving from an approach mainly based on the analysis of the urban/territorial factors and the real estate value, to a more integrate approach, in which not only the spatial and the financial aspects of the project are considered, but also the social implications and the environmental effects. Facing the new trends in the context of public participation at a European level, it is necessary to be more inclusive in the evaluation process, considering the use of specific tools enabling the involvement of the population in the decision process and to take the different opinions into account.

A fundamental aspect is that the planning context is usually very dynamic: the political relevance of items, alternatives or impacts may exhibit sudden change, hence it is important to conceive evaluation as a continuous activity that permanently takes place during the planning process. "It is noteworthy that evaluation processes often have a cyclic nature. By "cyclic nature" it is meant the possible adaptation of elements of the evaluation due to continuous consultations between the various parties involved in the planning process at hand. Such a learning process is a necessary condition to bridge the gap between technicians, researchers and planners" [27].

Evaluation takes place in all phases of decision making. The models facilitate dialogue [35, 47]. Lots of techniques and tools are available, depending on the phase in which the evaluation takes place (before, during or after the construction of the project). In the *ex ante* phase, the evaluation tools are necessary to support the formulation of the project, providing information both on the strategies (the objectives that the project is likely to pursue) and the visions (the actions that the project will implement in order to reach the objectives). The *in itinere* phase is mostly related to control whether the project meets the initial objectives, by emphasising the unexpected effects. Whereas in the *ex post* phase the evaluation process can help to make a final balance of the experience and to inform the local public administration and the population about the final results that the project attained.

Concerning the evaluation tools which can be used to measure the impact of transport policies and projects, in the *ex ante* phase monetary and non-monetary evaluations are used. A monetary evaluation is characterized by an attempt to measure all effects in monetary units, whereas a non-monetary evaluation utilizes a wide variety of measurement units. In particular four types of evaluation analysis are used in this context : the Cost-Benefits Analysis (CBA), the Cost-Effectiveness Analysis (CEA), the Discounted Cash Flows (DCF) and the MCDA (Table 6.1).

This chapter focuses on MCDA, for the following reasons:

- It takes all applicable units of measure into consideration. Therefore, it is possible to view all of the fundamental aspects of the operations and not just the monetary units (such as intervention costs or social opposition costs that bar the impact of noise pollution in the eventuality of developing new forms of urban settlement).
- It is possible to realise a project of such scale as the Eurocorridor. During the decision-making process of said project it is likely that at a certain moment the very purpose of the intervention is questioned. Consequently, MDCA is the ideal tool to manage this delicate stage.
- Often there is partial or missing information in the *ex ante* stage of an intervention of this scale and nature. Tools such as MCDA develop reasoning and interesting comparisons at macro-scenario level.
- The fact that the instruments are participatory in nature, which has been essential in building consensus.
- The ability of applying multicriteria techniques to actors at hand during the realisation processes of transport infrastructure on such a vast scale is very rare.

**Table 6.1** Main features of appraisal tools measuring the impact of transport policies and projects

| | CBA | CEA | DCF | MCDA |
|---|---|---|---|---|
| Description | Analysis of changes in social welfare over time associated with the intervention. It seeks to quantify all of the costs and benefits of a proposal in monetary terms, including items for which the market does not provide a satisfactory measure of economic value | Analysis that compares the costs of alternative ways of producing the same or similar outputs. It is generally used to assess the efficiency of certain technologies, programmes or policies in order to compare a number of alternatives | A valuation method used to estimate the attractiveness of an investment opportunity. It uses future free cash flow projections and discounts them to arrive at a present value | Analysis of the full range of aspects that are related to the project. It permits to integrate qualitative and quantitative information into a single assessment or output |
| Application | Predominantly road project evaluation but has been applied to demand management and technology policy options | Predominantly technology and alternative fuel policy options | Predominantly project-level | Predominantly project-level at a very early stage |
| Trend in use | Widely used, firmly embedded in project appraisal | Increasingly used as part of marginal abatement cost (MAC) curves but not widely formally integrated into policy appraisal | Widely used for both, private and increasingly public investment | Not widely used in practice but qualitative elements of MCDA increasingly used in project appraisal and for comparing scenario alternatives |
| Input | Monetary measure of changes in well-being, social discount rate | Costs | Costs and revenues, discount rate | Measurement of positive and negative impacts, utility functions, weights, etc |
| Output | Social rating convenience (Net Present Value—NPV, Internal Rate Return—IRR) | Cost—effectiveness ratio | Private rating convenience (NPV, IRR, Pay Back Period—PBP) | Decision ranking, rules, indicators, etc |

(continued)

**Table 6.1** (continued)

| | CBA | CEA | DCF | MCDA |
|---|---|---|---|---|
| Positive impact considered | Predominantly travel time savings and reduction in accidents and fatalities | Greenhouse gas—GHG emissions reduction | Predomintaly tolls or pre-sale contracts transportation services | Potentially all benefits |
| Stakeholder participation | Possible but not required | Possible but not required | Possible but not required | Formal part of process |
| Strengths | The result provided by the evaluation is easy to communicate—single value | The result provided by the evaluation is easy to communicate—single value | The result provided by the evaluation is easy to communicate—single value | The evaluations well represent the public decision making (conflicts' analysis, technical and political judgments etc.). The communicability depends on the technique used |
| Weaknesses | The monetization of externalities may be inaccurate or unacceptable | The estimation of all costs could be difficult | The evaluation ignore the public assessment (externalities) | The procedures have some uncertain results that are subject to high variation over time. High level of subjectivity |

*Source* Elaboration from Browne and Ryan [8]

As stated by Figueira et al. [13, p. 25], "MCDA is not just a collection of theories, methodologies, and techniques, but a specific perspective to deal with decision problems. Losing this perspective, even the most rigorous theoretical developments and applications of the most refined methodologies are at risk of being meaningless, because they miss an adequate consideration of the aims and of the role of MCDA".

There is a growing number of positive examples of using MCDA to support participatory and collaborative processes [14, 15, 20, 21, 26]. MCDA allow several criteria to be taken into account simultaneously in a complex situation and they are designed to help the Decision Makers (DMs) to integrate the different options into a prospective or retrospective framework [13, 36]. MCDA is a versatile and flexible approach to participatory processes allowing the stakeholders to engage and incorporate their values and knowledge into different phases of the planning process.

There are also some challenges and pitfalls in the use of MCDA, which may affect the quality and legitimacy of the outcome. As underlined by Marttunen et al. [26], these are often related to how well MCDA suits or is tailored to the question at hand and how professionally it is applied. Failure to identify the real nature of decision-making may place the resulting analysis at risk and greatly diminish the relevance of the results [28, 40]. At the same time, difficulties in reading output data, especially if these are numerical lists or matrixes, as well as the variety of the DMs' backgrounds can limit the process of data sharing and knowledge.

## 6.4 Choosing a Multicriteria Method

An illustrative and comprehensive list of the whole MCDA is beyond the scope of this chapter (please refer to [7, 13]), but what we would like to do is to provide some reflections that could help an analyst in choosing a method to be used in a decision context similar to the one here analysed.

It is possible to provide aid to someone who is struggling in the decision process by asking, said person as well as the analyst, a number of crucial questions [38].

Roy and Słowinski [39] provided a very interesting list of questions which, in their view, should be answered by an analyst before choosing the MCDA in any decision context. They suggest that the first essential question the analyst should start with reflecting on the best or even the only way of answering is the following: "Taking into account the context of the decision process, what type(s) of results is the method expected to bring, so as to allow elaboration of relevant answers to questions asked by the decision maker?". This question is fundamental because, depending on the decision context, the same type of results may not bring useful information able to guide the decision aiding process in the right way. The authors add, to the central one, five other key questions to choose the right method:

- Do the original performance scales have all the required properties for a rightful application of the considered method.
- Is it simple or hard (even impossible) to get preference information that the method requires.
- Should the part of imprecision, uncertainty or indetermination in the definition of performances be taken into account, and if so, in what way.
- Is the compensation of bad performances on some criteria by good ones on other criteria acceptable.
- Is it necessary to take into account some forms of interaction among criteria.

These questions are useful not only in a decision making process, but also in a decision aiding one. The decision making processes are the most widely used in order to conduct the Decision Maker (DM) to take a decision. In similar situations the DM has the necessary information to be able to conclude the process of finding a more satisfactory solution [38]. However, the decision processes are not always designed to come to a final decision, but could be concluded in the understanding of the problem, the description of the decision situation, the justification of the choices, discussion, persuasion etc. [37]. In such cases a decision aiding process is necessary. The presence of an analyst and the use of a decision support tool is essential in order to overcome the difficulties encountered during the decision process. In a decision aiding process there is no a real decision maker; the one who is asking for help (individual, organization, administration etc.) might not necessarily be interested in coming to a conclusion. Instead, he may be asking for a help because the decision situation is so complex, characterized by multiple stakeholders and decision variables, that requires an elaborate process of understanding before making a final decision [7].

It is important to add another concept to this framework, in order to comprehend how to develop a successful negotiation in decision making and decision aiding processes; the concept of "affordances". People do not interact with an object prior to or without perceiving what the object is good for: the perception of an object's utility could be called "affordance"[19]. Affordances are unique to the particular ways in which an actor, or a set of actors, perceives and uses the object. In the relational view, affordances of an artifact can change across different contexts even though its materiality does not [24, 43]. Norman [30] defines an affordance as something of both actual and perceived properties. When actual and perceived properties are combined, an affordance emerges as a relationship that holds between the object and the individual that is acting on the object.

Franco [17] explored how the models developed had the capacity to invoke different perspectives, knowledge and interests that were 'at stake' [9], and those involved were able to use that capacity to openly discuss and negotiate new meanings that led to new knowledge and significant changes within the partnership. He identified five model affordances:

- "*Tangibility*: the ability of a model to make its content visible and concrete. This affordance makes domain-relevant knowledge available and tangible, and a source of group discussion and negotiation.
- *Associability*: the ability of a model to relate its contents based on shared attributes. This affordance enables those involved to identify knowledge differences and dependencies.
- *Mutability*: the ability of a model to modify its contents on the spot. This affordance allows evolving knowledge-related discussions and negotiations to be reflected in the model incrementally.
- *Traceability*: the ability of a model to relate its contents temporally and structurally. This affordance offers opportunities for surveying and assembling knowledge-related discussions and negotiations.
- *Analysability*: the ability of a model to transform inputs into outputs. This affordance enables experimenting with different knowledge-related inputs, and calculating their impact".

The affordances outlined above seem very promising also in the context of transport planning [22], in particular when the use of multicriteria methods is associated to visualisations, as illustrated in the next paragraph.

### 6.4.1 The Code24 Experience

As already stated, construction of major transport infrastructures is currently widely debated in Europe. Top down approach to transport infrastructure planning is no longer viable and new approaches are needed: (1) negotiation rather than coercion, (2) agreement building rather than imposition. Within this perspective, many experiences of decision aiding processes have been experienced in the

**Fig. 6.1** Final workshop about the whole Corridor Genoa Rotterdam, ANP application (ETH Zurich, February 2014)

**Fig. 6.2** Final workshop about the whole Corridor Genoa Rotterdam, collaborative assessment (ETH Zurich, February 2014)

Code24 framework, with different evaluation characteristics and different level of participation (Figs. 6.1 and 6.2).

It has been a challenging project, whereby it was necessary to find evaluation tools adaptable to problems relating to different geographical scales (from urban until the European Union); it needed to be understandable and potentially "acceptable" by individuals with very diverse academic backgrounds, goals and cultures (as they came from different countries); be compatible with dynamic visualization tools; as well as to encourage the highest level of participation.

The Analytic Network Process (ANP [42]) was chosen for the following reasons.

- First, the type of results the ANP methodology assigned are numerical values to each potential action. This is particularly significant, because there is no vision without numbers. Even if the discussion about the Genoa-Rotterdam Corridor has been at "macro" levels on different occasions, it was important to give a concrete idea of feasibility of the proposed and compared solutions. The ANP, like other methods, offers the ranking of alternatives as a final result and, for this reason, provides a readable and immediately understandable result. Moreover the ANP methodology is able to produce a list of best actions to be further analysed by the people involved [6].
- Second, the original performance scale of the ANP method, the Saaty's fundamental scale of absolute numbers, has all the properties required for a correct application. Hence there is no need to transform or codify the original scale, which could cause the rise of arbitrary transformation that could, in turn, affect the process as a whole. It does not handle missing data but often it is possible "to circumvent" the problem, resetting the structure of the decision model.
- Third, the ANP may contribute to the construction and review of alternatives [1] as experienced during the project.
- Fourth, it is based on the assumption of the decomposition of a complex problem into simpler elements, systematizing the relationship among the nodes. Similarly, it uses the principle of pairwise comparison to simulate the process of the human mind [41, 42]. It helps to take into account the views of different actors, even with heterogeneous languages, allowing to develop participation, due to the focus groups where different actors and decision makers can deal directly with each other.
- Fifth, it is possible to combine the ANP with a new visualization tool [31], as illustrated in Chap. 11 (Masala and Pensa).
- Finally, the way in which the ANP is applied really coincides the iterative and interactive role which is increasingly required in an evaluation process [2, 23].

As for the applications of the method, please refer directly to the case studies (Chaps. 12, and 13, [2]).

The following Table 6.2 provides an overview and comparison of the attended workshops, with a synthetic description of the preparations phase (for more details please see Chap. 13 the case study of the entire corridor), with their main characteristics. On a side, note the workshops also increased and improved the researchers' learning processes.

Similar to research of Franco [16, 17], in these cases the affordances indicated on each application were perceived differently by the people involved. Generally, these affordances enabled a collaborative mode of interaction to be adopted, in

**Table 6.2** Overview of the evaluations workshops experienced in the project CODE24

| Time | Activity | ANP structure | Number of questions | Weight elicitation | Preparing the focus group | Stakelhoders | Affordances |
|---|---|---|---|---|---|---|---|
| February 2011 | Ranking of the possible effects of a delay in the construction of the Corridor 24 in Italy | Simple network | 31 | Arithmetic Average | Information material | 8 (experts internal to the CODE24 project, all Italian) | Tangibility |
| July 2011 | Evaluation of three transport scenarios for the area of Bellinzona (CH) | BOCR model | 49 | Arithmetic Average | Information material | 12 (mostly experts internal to the CODE24 project. Italian, German and Swiss stakeholders) | Tangibility |
| December 2011 | Finding the best development strategy for the area of Frankfurt/ Mannheim (Germany)—Test Workshop | BOCR network | 100 | "Majority method" | Information material | 16 (partly the experts; small number of real actors—representatives of the Public Administrations—and some observers. Italian, German, Dutch and Swiss stakeholders) | Tangibility<br>Editability<br>Associability |
| March 2012 | Finding the best development strategy for the area of Frankfurt/ Mannheim (Germany)—Workshop | BC network | 35 | The questionnaire form was partly filled in under the guidance of MCDA expert in the workshop and partly at home. The elicitation of the weight was done by the "Majority method". | Information material in English and German; Online ANP questionnaires in English and German, selection of 10 key questions | 10 (experts internal and external to the project Code24, real DMs and citizen representatives. Italian, German, Dutch and Swiss stakeholders) | Tangibility<br>Editability<br>Associability<br>Mutability<br>Traceability (partially) |

(continued)

**Table 6.2** (continued)

| Time | Activity | ANP structure | Number of questions | Weight elicitation | Preparing the focus group | Stakelhoders | Affordances |
|---|---|---|---|---|---|---|---|
| September 2012 | Finding the best development strategy for the Corridor 24—Test Workshop | Complex strategic network—BC network | 100 | “Majority method” | Information material in English; Online ANP questionnaires in English, selection of 32 key questions | 9 (experts internal to the project Code24, real DMs. Italian, German, Dutch, French and Swiss stakeholders) | Tangibility<br>Editability<br>Mutability<br>Traceability (partially) |
| February 2013 | Finding the best development strategy for the Corridor 24<br>Workshop | Simple network | 19 | “Majority method” | Information material in English; Online ANP questionnaires in English | 10 (experts internal and external to the project Code24, real DMs. Italian, German, Dutch, French and Swiss stakeholders) | Tangibility<br>Editability<br>Associability<br>Mutability<br>Analysibility<br>Traceability (partially) |

order to design much wider perspectives and different possibilities to transform the analysed territories.

At the end of this long process of experimentation of evaluation workshops on the theme of the euro-corridors, the following conclusions have been reached:

- In order to obtain the trust and effective involvement of the participants it was crucial to involve the stakeholders in the definition/redefinition of the objective of the decision-making process, to design of the alternative scenarios and to identify of clusters and nodes. The process was important because "there is usually no shared understanding of terms like mission, vision, goal, objective" [33] in [12].
- The use of real-time excel sheets (for the aggregation of the weights assigned during the discussion) and dynamic maps (to represent the distribution of the expected effects in the area) has increased exponentially, respectively, the tangibility and traceability, as well as the mutability for the presence of both the tools.
- The analysability was increased by simplifying the adopted ANP network and by reducing the level of abstractness of the visualizations.
- By sending the ANP questionnaire in advance it allowed the participants to compile it prior to the workshop which has ensured a greater associability to the project.

## 6.5 Conclusions

The purpose of this chapter has been to set out the background to territorial conflict with particular reference to the realisation of mayor transport infrastructure, and to introduce the idea of MCDA in combination with visualisation tools to tackle these types of decision problems.

Drawing on recent developments in the area of MCDA, it has been discussed the issue that the valuation methods to support decision-making problems are numerous. The choice of one rather than another depends not only on the type of expected results and on a series of elements such as the preference information, the management of uncertainty, the desire to use a compensatory method and the interaction between criteria, but also depends on their affordances, i.e. the possibilities for action they offer to those involved. The many and varied experiences of using multicriteria as tools to support decision aiding processes in a European project have stimulated general reflections with the possibility of increasing the affordances, varying the degree that was crucial to enable or constrain (model-supported) meaning negotiations and new knowledge creation [17].

## References

1. Abastante F, Lami IM (2013) An analytical model to evaluate a large scale urban design competition. GEAM 2(139):27–36
2. Abastante F, Bottero M, Greco S, Lami IM (2013) Dominance-based rough set approach and analytic network process for assessing urban transformation scenarios. Int J Multicriteria Decis Mak 3(2/3):212–234

3. Belton V, Stewart TS (2002) Multiple criteria decision analysis: an integrated approach. Kluwer Academic Publishers, Massachusetts
4. Bobbio L (ed) (2004) A piú voci. Edizioni Scientifiche Italiane, Roma
5. Bobbio L (2011) Conflitti territoriali: sei interpretazioni. Tema 4(4):79–88
6. Bottero M, Lami IM (2010) Analytic network process and sustainable mobility: an application for the assessment of different scenarios. J Urbanism 3(3):275–293
7. Bouyssou D, Marchant T, Pirlot M, Tsoukiàs A, Vincke P (2006) Evaluation and decision models with multiple criteria: stepping stones for the analyst. Springer, Boston
8. Browne D, Ryan L (2011) Comparative analysis of evaluation techniques for transport policies. Environ Impact Assess Rev 31:226–233
9. Carlile RP, Rebentisch ES (2003) Into the black box: the knowledge transformation cycle. Manage Sci 49(9):1180–1195
10. Castells M (2000) The rise of the network society, 2nd edn. Blackwell, Oxford
11. Dente B (2014) Understanding policy decisions. Applied science and technology, vol 6. Springer, Boston (forthcoming)
12. Eden C, Ackermann F (2013) Problem Structuring: on nature of, and reaching agreement about goals. EURO J Decis Process 1:7–28
13. Figueira J, Greco S, Ehrgott M (2005) Multiple criteria decision analysis: state of the art surveys. Springer, Boston
14. Franco LA, Montibeller G (2010) Facilitated modeling in operational research. Eur J Oper Res 205(3):489–500
15. Franco LA, Rouwette E (2011) Decision development in facilitated modeling workshops. Eur J Oper Res 212:164–178
16. Franco LA, Lord E (2011) Understanding multi-methodology: evaluating the perceived impact of mixing methods for group budgetary decisions. Omega 39:362–372
17. Franco LA (2013) Rethinking soft OR interventions: models as boundary objects. Eur J Oper Res 231:720–733
18. Friend J, Hickling A (2005) Planning under pressure: the strategic choice approach, 3rd edn. Elsevier, Amsterdam
19. Gibson JJ (1986) The ecological approach to visual perception. Lawrence Erlbaum Associates, Hillsdale, NJ
20. Hostmann M, Borsuk M, Reichert P, Truffer B (2005) Stakeholder values in decision support for river rehabilitation. Archiv für Hydrobiologie Supplement 155(1–4):491–505
21. Kiker GA, Bridges TS, Varghese A, Seager TP, Linkov I (2005) Application of multicriteria decision analysis in environmental decision making. Integr Environ Assess Manage 1(2):95–108
22. Lami IM, Günther F, Tosoni I, Abastante F, Franco LA (2013) Facilitated modelling workshops to support Corridor 24 development. In: 26th European conference on operational research, Rome, 1–4 July 2013
23. Lami IM, Abastante F (2014) Decision making for urban solid waste treatment in the context of territorial conflict: can the Analytic Network Process help? Land Use Policy 41:11–20
24. Leonardi PM, Bailey DE (2008) Transformational technologies and the creation of new work practices: making implicit knowledge explicit in task-based offshoring. MIS Q 32:411–436
25. Mclymonta K, O'hareb P (2008) "We're not NIMBYs!" Contrasting local protest groups with idealised conceptions of sustainable communities. Local environment. Int J Justice Sustainability 13(4): 321–335
26. Marttunen M, Mustajoki J, Dufva M Karjalainen TP (2013) How to design and realize participation of stakeholders in MCDA processes? A framework for selecting an appropriate approach. EURO J Decis Process. doi: 10.1007/s40070-013-0016-3
27. Munda G, Nijkamp P, Rietveld P (1994) Qualitative multicriteria evaluation for environmental management. Ecol Econ 10:97–112
28. Munda G (2008) Social multi-criteria evaluation for a sustainable economy. Springer, Berlin

29. Næss P (2006) Cost-benefit analyses of transportation investments: neither critical nor realistic. J Crit Realism 5(1):32–60
30. Norman DA (1999) Affordance, conventions, and design. Interactions 6:38–43
31. Pensa S, Masala E, Lami IM, Rosa A (2014) Seeing is knowing: data exploration as a support to planning. Civ Eng Spec Issue 167(CE5):3–8
32. Pichat P (1995) La gestion des déchets. Hérissey, Evreux
33. Phillips L (1990) Decision analysis for group decision support. In: Eden C, Radford J (eds) Tackling strategic problems. Sage, London, pp 142–150
34. Roscelli R (ed) (2005) Misurare nell'incertezza. Celid, Torino
35. Rosenhead J, Mingers J (eds) (2001) Rational analysis for a problematic word Revised. Wiley, Chichester
36. Roy B, Bouyssou D (1993) Aide Multicritère à la Decision: Méthodes et Cas. Economica, Paris
37. Roy B (1994) On operational research and decision aid. Eur J Oper Res 73:23–26
38. Roy B (1996) Multicriteria methodology for decision aiding. Kluwer, Dordrecht
39. Roy B, Słowinski R (2013) Questions guiding the choice of a multicriteria decision aiding method. EURO J Decis Process 1:69–97
40. Salgado P, Quintana CS, Pereira AG, del Moral Ituarte L, Mateos BP (2009) Participative multi-criteria analysis for the evaluation of water governance alternatives: a case in the Costa del Sol (Málaga). Ecol Econ 68:990–1005
41. Saaty TL (1980) The analytic hierarchy process, planning, priority setting, resource allocation. McGraw-Hill, New York
42. Saaty TL (2005) Theory and applications of the analytic network process. RWS Publications, Pittsburgh
43. Treem JW, Leonardi PM (2012) Social media use in organizations: exploring the affordances of visibility, editability, persistence, and association. Commun Yearb 36:143–189
44. Tsoukiàs A, Montibeller G, Lucentini G, Belton V (2013) Policy analytics: an agenda for research and practice. EURO J Decis Process 1:115–134
45. Van der Horst D (2007) NIMBY or not? Exploring the relevance of location and the politics of voiced opinions in renewable energy siting controversies. Energy Policy 35(5):2705–2714
46. Wester-Herber M (2004) Underlying concerns in land-use conflicts—the role of place-identity in risk perception. Environ Sci Policy 7:109–116
47. White L (2009) Understanding problem structuring methods interventions. Eur J Oper Res 99(3):823–833
48. Zeleny M (1982) Multiple criteria decision naking. McGraw-Hill Book Company, New York

# Transport Policy and Regional Development: The Economic Impact of Regional Accessibility on Economic Sectors

7

Hansjörg Drewello

**Abstract**

Regional accessibility is an important location factor, which enhance the region and, therefore, promote economic growth. It is the virtual aim of transport policy, on European as well as on national level, to strengthen regional accessibility. But available resources to do so are limited. Therefore, new investment and enlargement in transport infrastructure should be based on efficiency considerations. This article focuses on the economic impact of regional accessibility and in different economic sectors in order to prepare a methodology of planning for efficient transport infrastructure investment along the corridor Rotterdam-Genoa. The approach takes into consideration that the output of economic sectors depends differently on logistic services. If we consider different regional concentrations of specific sectorial activity, we can assume different economic impacts of transport infrastructure investment on different regions or, in other words, higher 'costs of non-doing' if bottlenecks in transport infrastructure persist. Therefore, the link between transport costs and regional accessibility in transport infrastructure is analyzed. The results of an empirical study which surveys the correlation between regional accessibility and regional sector output will be described.

H. Drewello (✉)
University of Applied Sciences Kehl (D), Kehl, Germany
e-mail: drewello@hs-kehl.de

I. M. Lami (ed.), *Analytical Decision-Making Methods for Evaluating Sustainable Transport in European Corridors*, Sxi 11, DOI: 10.1007/978-3-319-04786-7_7,

## 7.1 Introduction

The positive impact of transport infrastructure on the economic development of nations is not a subject of dispute in theory. In macroeconomics, investment in this kind of infrastructure increases the stock of public capital, which is an important input factor in the production of total national output.

On a regional level the correlation between infrastructure investment and economic growth is less evident. In Location Theory, the availability of transport infrastructure and regional accessibility are important location factors which enhance the region and, therefore, promote economic growth. Other models of Regional Science or Economic Geography assume that a high level of transport infrastructure decreases transport costs. In these models, transport costs have a major impact on the agglomeration of economic activity. Public investment in infrastructures can decrease regional economic growth if industries decide to move production to more important agglomerations. Nevertheless, European transport policy aims to enforce regional cohesion by facilitating an above-average increase in economic growth in weaker regions.

The aim of this article is to focus on the economic impact of regional accessibility and in different economic sectors in order to prepare a methodology of planning for efficient transport infrastructure investment along the corridor Rotterdam-Genoa. This approach takes into consideration that the output of economic sectors depends differently on logistic services. If we consider different regional concentrations of specific sectorial activity, we can assume different economic impacts of transport infrastructure investment on different regions or, in other words, higher 'costs of non-doing' if bottlenecks in transport infrastructure persist.

Transport costs are of particular interest. An important issue, therefore, is how economic theories like macro- and microeconomics, international trade theory or geographical economics deal with transport costs. A brief illustration of the Code24 transport model is given, which is based on the cost of freight transport. The model allows for a calculation of regional accessibility in terms of freight transport along the Rotterdam-Genoa corridor from different aspects. A further issue of this article is to analyze the link between transport costs and regional accessibility in transport infrastructure. Finally, the results of an empirical study which surveys the correlation between regional accessibility and regional sector output will be described.

## 7.2 European Transport Policy on Rail: A Brief Overview

The aim of current European transport policy is to strengthen rail transport by opening up rail markets to greater competition, promoting technical standardization between rail systems and modernizing Europe's rail infrastructure [1]. The new EU infrastructure policy triples EU financing to EUR 26 bn for transport for the period 2014–2020. At the same time it refocuses transport financing on a tightly defined new core network. It will remove bottlenecks, upgrade infrastructure and streamline cross border transport operations.

This policy is also part of the European cohesion policy (e.g. [2]). Better transport conditions in this sense aims to remove economic, social and territorial disparities across the EU, restructure declining industrial areas and diversify rural areas that have declining agriculture. The title of the European White Paper on transport describes the overall objective of this policy: European transport policy aims at a resource efficient transport system, especially along important transport corridors.

This European transport policy is sometimes in conflict with national policy alignment. In Germany e.g. the federal states and the infrastructure manager Deutsche Bahn Netz propose rail transport projects for each German Federal Transport Infrastructure Plan (German: Bundesverkehrswegeplan; [3]). The choice of those projects is not based on efficiency considerations. Political considerations in the federal states aim at regional balanced investment. Deutsche Bahn Netz is also part of the vertically integrated national incumbent holding of the German railway system. Market power in the infrastructure market can be transferred to potentially competitive markets and used for discrimination of competitors.

The very same problem appears when national transport policy dictates which project of the national plan has to be realized first. Indeed, each national transport project is checked by a cost-benefit-analysis [4]. Projects are only approved if benefits exceed costs.

But this doesn't automatically mean that the evaluated project is efficient. Furthermore, cost-benefit analysis is susceptible to political influence and more importantly, the German planning process doesn't foresee an efficiency-based prioritization of those national transport projects. The new planning process for the period from 2015 until 2030 includes only a level three prioritization which is based especially on considerations of traffic flow [3].

## 7.3 Economics and Transport Policy

Economics is a study of how people and society end up choosing to employ scarce productive resources that could have alternative uses. It studies the production of various commodities over time and their distribution for consumption, now or in future, among various groups in society in order to maximize social welfare [5]. This definition by the famous economist and Nobel Prize laureate Paul Samuelson fits very well with regard to current problems in European transport development. Transport infrastructure is such a scarce productive resource. The German Federal Transport Infrastructure Plan is a very good example. The Plan has suffered from a chronic shortage of financing right from the start. The current plan, which was published in March 2003, foresees an investment of EUR 150 bn in transport infrastructure in the period from 2001 to 2015. This corresponds roughly to a yearly spending of EUR 10 bn on rail, road and barge in [6]. On average, EUR 3.5 bn should be spent on rail each year for maintenance and new investment [7].

According to nine important German economic associations, the need for public investment in transport infrastructure exceeds the real investment to a large degree. Every year since 2001, with the exception of 2009 and 2010, public spending has been lower than EUR 10 bn. The associations estimate the need at EUR 14 bn [8]. This is the classic economic dilemma in societies where unlimited needs encounter invariably limited resources.

The mismatch between need and resource becomes clearer when one considers the European Rotterdam-Genoa transport corridor. The project group of Code24 estimates the total costs of all suggested infrastructure projects to be EUR 35 bn. These projects have been highlighted in different workshops and regional round tables along the corridor by regional stakeholders. Even if we refer only to the "most important infrastructure projects," investments of about EUR 19 bn seem to be necessary [9]. But the total need in Netherlands, Germany, Switzerland and Italy for railway infrastructure is about EUR 210 bn. A planning horizon of at least 40 years seems to be realistic for the realization of the extension of the corridor.

Therefore, it is necessary—from an economic point of view—to realize first those infrastructure projects which lead to highest economic benefit. In general, these will be projects which are able to increase infrastructure capacity in regions where infrastructure bottlenecks lead to higher transport costs due to an increase in transport or travel time and a decrease in reliability. Improvements in transport infrastructure decrease transport costs by shortening transport distances or increasing capacity for transportation.

Economic theory deals differently with transport costs. In macroeconomics, the importance of transportation for a whole economy is analyzed. Transport costs are linked to a level of output, employment, consumption and income within a national economy (e.g. [10]). For example, transport costs are part of the aggregated cost function of a national economy. A decline in transport costs will lead ceteris paribus to an increase in output in the economy. But transport investments are supposed to have declining marginal returns. Initial infrastructure investments have a high return since they provide new mobility options and so reduce transport costs. The more the system is developed, the more likely additional investment would result in lower returns.

In microeconomics, the importance of transportation for specific markets of the economy is analyzed. Transport costs are linked to producer, consumer and production costs. The importance of specific transport activities and infrastructure can thus be assessed for each sector of the economy.

There are also significant negative impacts to consider which influence individuals, markets or society in general. Those negative external effects can be congestion in the transport system, accidents, which tend to be proportional to the intensity of the use of transport infrastructures, air pollution and noise, or land usage because transport infrastructure takes up large amounts of space.

In international economics, the impact of transport costs on the volume and nature of international trade is of more interest. Transport costs are one factor, amongst many others, that shapes trade patterns. They are in turn determined by underlying variables such as distance, geography, infrastructure quality, trade facilitation measures, fuel costs and transport technology. Integrated into the classical Ricardian model of comparative advantage, transport costs reduce or prevent trade. Certain goods may not be traded internationally like haircuts or other skilled manual work [11]. Transport costs also influence choice of mode (e.g. [12]), the commodity composition of trade [13] and the organization of production, particularly as 'just-in-time' methods, outsourcing and centralized distribution strategies are found all over the world [14].

Furthermore, Hummels and Skiba argue that transport costs lead to the export of higher-quality products [15]. They explain that transport cost per unit ceteris paribus falls proportionately as the price of the good rises. Transport costs are not to be considered as absolute costs, but rather in relation to the transported value.

In geographical economics, many of the above-mentioned relationships between transport cost and economic development are singled out. The important role of transport infrastructure for spatial development and, hence, transport cost implies that areas with better access to the locations of input materials and markets will be, ceteris paribus, more productive and competitive than more isolated areas. Regional agglomeration theories are early attempts to explain the concentration of economic activity in geographical space. Alfred Weber formulated a theory of industrial location in which an industry is located where the transport costs of raw materials and final product is a minimum [16]. The central place theory of Christaller explains number, size and location of human settlements in urban systems [17]. In this theory, transport cost is proportional to distance. Lösch modified Christaller's theory by creating an ideal consumer landscape where the need to travel for any good was minimized [18].

The relation between infrastructure endowment (or transport cost) and regional development can be modelled in a regional production function (e.g. [19, 20] or in functions where simple infrastructure impact is substituted by more complex accessibility indicators [21].

## 7.4 Bottlenecks, Regional Accessibility and Transport Costs

Bottlenecks in the transport system occur particularly in times of transport growth. In public debate, it is sometimes unclear what is meant by a bottleneck in transport infrastructure. In many of these political statements, a lack of capacity is specified, usually combined with the projection of increasing freight transport. Holzhey determines capacity bottlenecks by calculating the potential maximum of freight train paths per day for a corridor and comparing that with future needs [22].

As early as 1996 Rothengatter recognized that technical capacity was not a sufficient measure to identify major deficiencies in railway networks. For him, insufficient service levels of railway companies were at that time more important than technical bottlenecks [23]. Cipolina and Ghiara distinguish four different categories of bottlenecks in freight transportation: infrastructural, organizational, technical and bureaucratic [24].

One important target of the Code24 project is to better understand bottlenecks in logistics and their effects on the freight transport corridor Rotterdam-Genoa. In order to do so, an international and interdisciplinary expert group within the project (planers, architects, engineers, logisticians and economists) developed a common definition in 2011 by means of a structured brainstorming process, called Metaplan.

> Bottlenecks always represent additional costs to logistic services by hindering them. They can be observed on a politico-legal, organizational or physical level. Such a bottleneck can be national or transborder [25].

What are the economic effects of these kinds of bottlenecks? They decrease regional accessibility. Good accessibility of regions improves their competitive position. Mobility and accessibility are key prerequisites for economic development of all regions of the EU. Another definition is that "accessibility indicators describe the location of an area with respect to opportunities, activities or assets existing in other areas and in the area itself, where 'area' may be a region, a city or a corridor" [26].

Accessibility is the main product of the transport system. It determines the locational advantage of an area relative to other areas. With accessibility indicators, one can measure the benefits for households and firms which profit from the existence and use of transport infrastructure.

To measure regional accessibility, various indicators were developed in the past. The basic principle of these indicators is relatively simple. It refers to Newton's law of universal gravitation, which states that any two bodies in the universe attract each other with a force that is directly proportional to the product of their masses and inversely proportional to the square of the distance between them. It was adapted by Stewart for use in regional economics [27]. The idea of those indicators is that the accessibility of a region is directly proportional to the attractiveness or size of the region, and indirectly proportional to distance, travel time or cost. Hence, regional accessibility is a function of regional attractiveness and transport cost

$$A_i = \sum_j g(W_j) \cdot f(c_{ij}) \tag{7.1}$$

where $A_i$ is the accessibility of a region i, $W_j$ the attractiveness or size of a region j, and $c_{ij}$ the general cost to reach region j from region i. $g(W_j)$ is a function of regional attractiveness, e.g. regional GDP, population or employment rate, and $f(c_{ij})$ of space resistance, e.g. transport cost or time between regions [28]: the more attractions in region j and the more easily reachable from region i, the higher the accessibility of region i.

Various types of accessibility indicators can be created by specifying the functions g(Wj) and f(cij) differently (linear, or non-linear). Wegener et al. describe functions of travel time or weighted travel time, of daily accessibility, potential accessibility, multimodal or intermodal accessibility [28].

Space resistance is measured basically in terms of transport cost, travel time or quality of transport. From an economic point of view, travel time and quality of transport can be included in travel cost, too. A long freight transport time increases transport costs because capital (vehicles and goods) works less efficiently than in shorter transport times. It is the same for passenger transport, where longer transport time creates higher costs because people cannot use the time alternatively. If quality of transport represents punctuality, a bad quality (declining punctuality) in the transport system creates higher costs because of increased uncertainty. The Code24 model contains all this information for regions along the Rotterdam-Genoa corridor (see Sect. 7.3) and includes it in transport cost.

## 7.5 Accessibility and Regional Sector Economy—a Correlation Analysis

### 7.5.1 Theoretical Considerations

Microeconomic theory suggests that different markets are differently affected by transport costs. Therefore, certain industries might suffer more from an increase in those costs than others. Available data is not very recent. The US Department of Transportation gives some evidence to this theory [29]. It has developed a set of transportation satellite accounts which seeks to establish the full picture with regard to the scale of transportation costs in the US economy in 1997. The researchers found that the economic sector "construction" was the most intensive user of transport at 14.6 cents transport cost per dollar output. The information and financial services sectors were considerably lower at 1.2 and 1.7 % respectively. For Europe, Meyer-Rühle et al. found further evidence for this theory in a study which covers the EU27 countries in 2000 [29]. The highest share of transport cost in output was found for wholesale and commission trade (21 %). The share of transport inputs in retail trade output was 7 %. Manufacturing varied between 5 % (basic metals) and 2 % (communication).

When a firm relocates, transport issues are not in general a first-order consideration, especially if these costs are only a small part of total production costs. However, these costs may be important in several sectors if the ratio of transportation costs to production costs is high or if accessibility to customers may influence the performance of a company [30].

If companies act in this way, a higher concentration of transportation-sensitive sectors should relocate in regions with well-developed accessibility. A correlation analysis shall test this theoretical assumption for regions along the Rotterdam-Genoa transport corridor.

### 7.5.2 Data Sources

To carry out the analysis of regional accessibility, a dataset was compiled from Eurostat data which, in addition to information about population, area and gross added value of the corresponding regions, also contains information about the regional gross added value of the individual sectors of the economy. There is also data on the number of employees in the various sectors.

The regions examined are organized in accordance with the NUTS classification system (Nomenclature des unités territoriales statistiques). There is data for Germany, France, Belgium, the Netherlands, Austria, Liechtenstein and Luxemburg at the NUTS 3 level. For Switzerland and Italy, there is only some data at the NUTS 2 level, which however can serve to make a plausible estimate for the NUTS 3 level. The different methods used in the classification systems of the individual countries mean that not all variables are available for each country, especially with regard to the classification of economic sectors according to NACE Rev. 2. As a result, the correlation between the calculated indicator and the added value of the individual sectors cannot be determined for each country and each sector. In particular, for the financial sector as well as for the information and communications sector, there is no specific information about added value for Germany, for these sectors are a subset of a broader category of economic activity and as such are accounted for in the statistics of the latter.

The data necessary for calculating regional space resistance in the accessibility indicators derives from the CODE24 model. This data concerns without exception freight transport. Transport costs between regions are calculated by using the minimum costs of intermodal relations: road, rail and/or barge.

Germany, France, Italy, Switzerland, Luxemburg, Liechtenstein, Belgium and the Netherlands are defined as regions of the corridor or of its direct catchment area. Austria is also viewed in high resolution due to its proximity to the corridor countries. The indicator is calculated for these regions. Geospatial data used to create maps were taken from the geographic information system Eurostat/GISCO.

Data on the non-EU countries, which are needed to calculate the indicator, were taken from the CIA World Factbook[31] and the World Economic Outlook Database of the IMF[32]. Despite their generally high reliability, there are many gaps in the Eurostat data. Thus for practical reasons, only data for the year 2009 are used. They have the least amount of gaps. This simplification is justified because the quality of the transport network over the entire corridor over time varies only slightly. Missing GDP data for regions of the corridor were imputed by multiplying the per capita GDP of the parent NUTS 2 region with the average population of the region, provided this information was included in the dataset. If the population figure was also missing, then the arithmetic mean of the GDP for the rest of the NUTS 3 regions of the same country was used.

### 7.5.3 The Code24 Transport Model

The CODE24 transport model aims at an assessment of the accessibility of a region with regard to logistics and transport services. The model is built up within VISUM as the software environment [33]. By means of the VISUM software environment, all available information about regions, transport networks, terminals and "schedules" (services), costs as well as transport or monetary flows is integrated into the model. This information is used within routing schemes. The model tries to find the shortest or most efficient path, e.g. with regard to transport costs. The analysis leads to cost matrices (COMAs) containing standardized transport costs for all origin/destination pairs within the CODE24 transport corridor and for selected origins/destinations outside the corridor. These COMAs can be used for further analysis inside (e.g. combination with monetary or transport flows, map-based displays) and outside the VISUM environment in matrix calculations for specific case studies [34].

Within this routing/shortest path calculation, quite different types of information are considered such as: networks, terminals, services, network-based cost as derived by use of specific cost-calculation schemes, waiting and transshipment times, transshipment cost, insurance cost and time losses.

The cost calculation can be done for different transport segments such as containers, bulk goods, mineral oil products or automotive. Different kinds of goods show quite different transport and cost structures. This analysis focuses on transport cost for containers.

### 7.5.4 Specifications of the Accessibility Indicator

To account for agglomeration effects and thus to take into consideration the stronger pull of very strong economic regions in the model, the additional parameter $\alpha$ is greater than one is introduced into the general equation of regional accessibility (1) as an exponent of the activity function $g(W_j)$—in this case, gross domestic product. To give more weight to distances, the space resistance function f(cij) is designed with the transport costs as a negative exponential function. A high value for parameter $\beta$ means that nearby targets are more heavily weighted. The resulting function can also be called a potential function of accessibility [28, 35]. It follows that

$$A_i = \sum_j BIP_j^{\alpha} \cdot \exp\left(-\beta \cdot c_{ij}\right). \tag{7.2}$$

The specifications of parameters $\alpha$ and $\beta$ in Table 7.1 were chosen to weight regional attractiveness and distance factors of the indicator [36].

Indicators $I_{1.1}$, $I_{2.1}$ and $I_{3.1}$ involve a high weighting for agglomeration effects, and a lower weighting for (cost) distances. Indicators $I_{1.2}$, $I_{2.2}$ and $I_{3.2}$ however, weights agglomeration effects lower while showing just as high a weighting of distance as $I_{1.1}$. Indicators $I_{1.1}$ and $I_{1.2}$ contain only transport cost on road, $I_{2.1}$ and $I_{2.2}$ on road and rail and $I_{3.1}$ and $I_{3.2}$ on road, rail and barge.

**Table 7.1** Overview of calculated indicators

| Indicator | $\alpha$ | $\beta$ | Regional attractiveness | Cost function based on mode of transport |
|---|---|---|---|---|
| $I_{1.1}$ | 2 | 1 | GDP | Road |
| $I_{1.2}$ | 1 | 1 | GDP | Road |
| $I_{2.1}$ | 2 | 1 | GDP | Road and rail |
| $I_{2.2}$ | 1 | 1 | GDP | Road and rail |
| $I_{3.1}$ | 2 | 1 | GDP | Road, rail and barge |
| $I_{3.2}$ | 1 | 1 | GDP | Road, rail and barge |

## 7.6 Regional Accessibility Along the Rotterdam-Genoa Corridor

The illustration of the indicator $I_{3.1}$ (Fig. 7.1) shows a clear north-south divide in terms of accessibility. The regions of greater Paris, the Ruhr and the Netherlands are the most accessible, and Italy (with the exception of the area around Rome) and southern France are the least accessible. The Alps are clearly a barrier to the transport of goods. The concentric circular pattern of accessibility intervals around the centers of Rotterdam and Antwerp is very noticeable.

Figure 7.1 illustrates the influence of the river Rhine to the accessibility of regions. Rings of decreasing accessibility centered around the North Sea ports of Rotterdam and Antwerp are still visible. The potential use of barge transportation which is included in the calculation of indicator $I_{1.3}$ increases the accessibility of regions near the river Rhine and illustrates the limitation of the corridor until Basel (Fig. 7.2).

The illustrations of indicators $I_{1.1}$ (road) and $I_{2.2}$ (road and rail) present similar pictures as $I_{3.2}$, with rings centered around the North Sea ports of Rotterdam and Antwerp (Figs. 7.3 and 7.4). Here again the Alps can be seen as a transport barrier. However, these illustrations clearly show that the Rotterdam-Genoa corridor is in the zone of highest accessibility even if accessibility in the corridor is not homogenous.

## 7.7 Correlations of Regional GDP with Regional Accessibility

The correlations of the indicators differ significantly between the specifications. The following table shows the relationship between the individual indicator specifications and the added value of the different sectors (Table 7.2).

All accessibility indicators are negatively correlated with the added value of the agricultural sector, but mostly positively with all other variables. The strongest correlations are those with the added value of information, trade, services finance and construction sector.

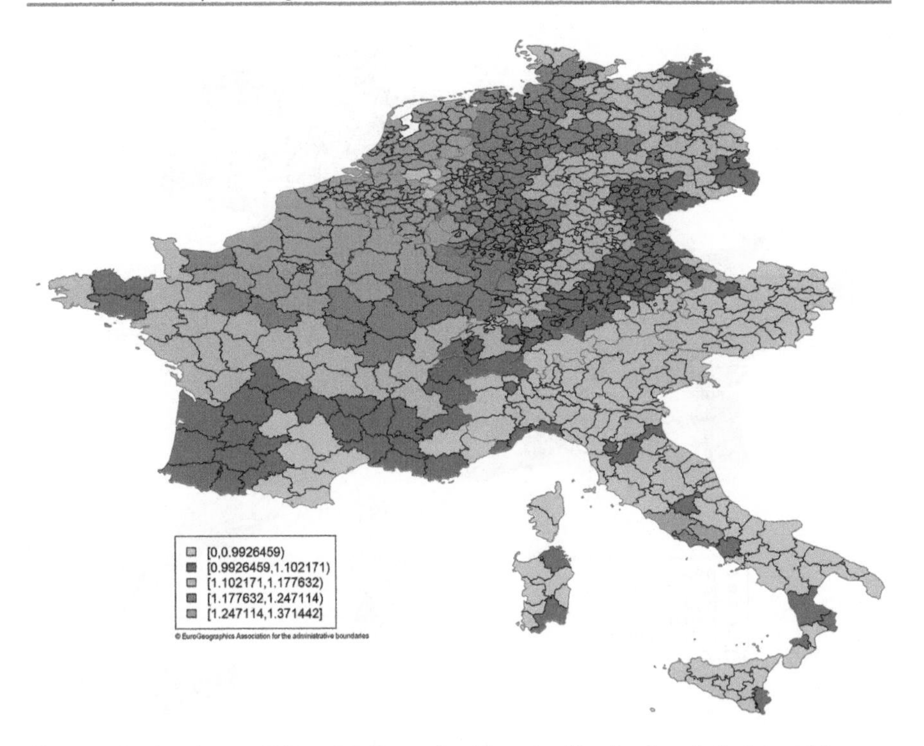


**Fig. 7.1** Regional accessibility of Indicator $I_{3.1}$ (data classification in quintile intervals)

Indicators representing road transportation are generally higher correlated with added value of economic sectors than indicators representing rail and road. The two indicators representing all modes of transport ($I_{3.1}$ and $I_{3.2}$) are slightly positively correlated with all sectors except agriculture. Probably this result reflects the relative importance of road transportation for those sectors in relation to rail and barge. Especially the sectors of trade, services and construction seem to be more productive in regions which are well developed in terms of transport infrastructure. The GDP of the sectors of information and finance is highly correlated with the accessibility indicators of road and even of road and rail. Because these sectors do not transport any container, the interpretation is more difficult. Probably a high correlation between regional accessibility and other important location factors exists.

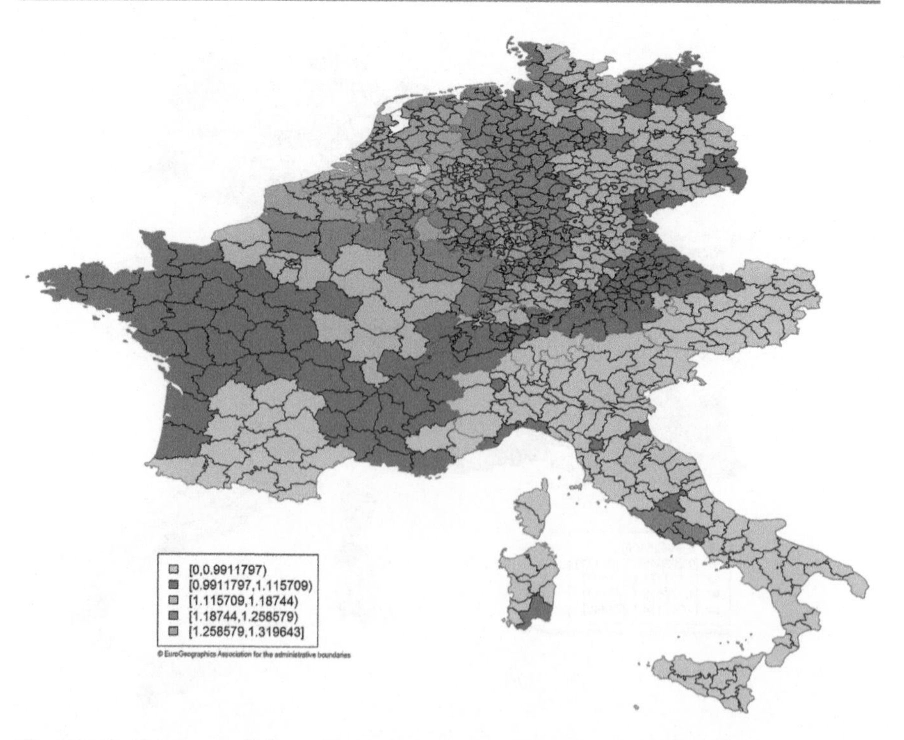


**Fig. 7.2** Regional accessibility of Indicator $I_{3.2}$ (data classification in quintile intervals)

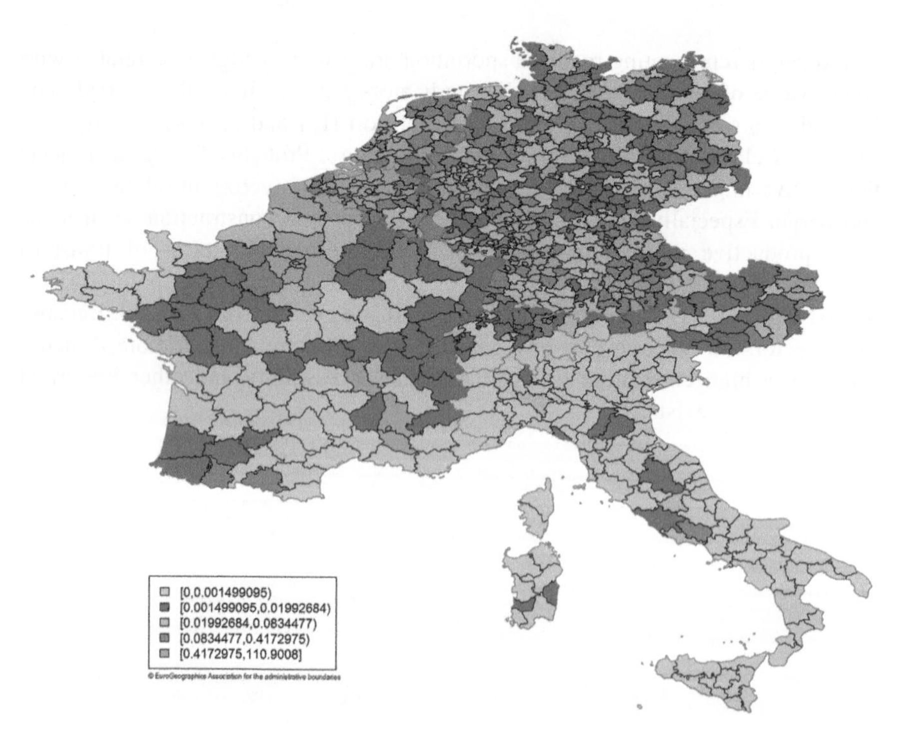


**Fig. 7.3** Regional accessibility of Indicator $I_{1.1}$ (data classification in quintile intervals)

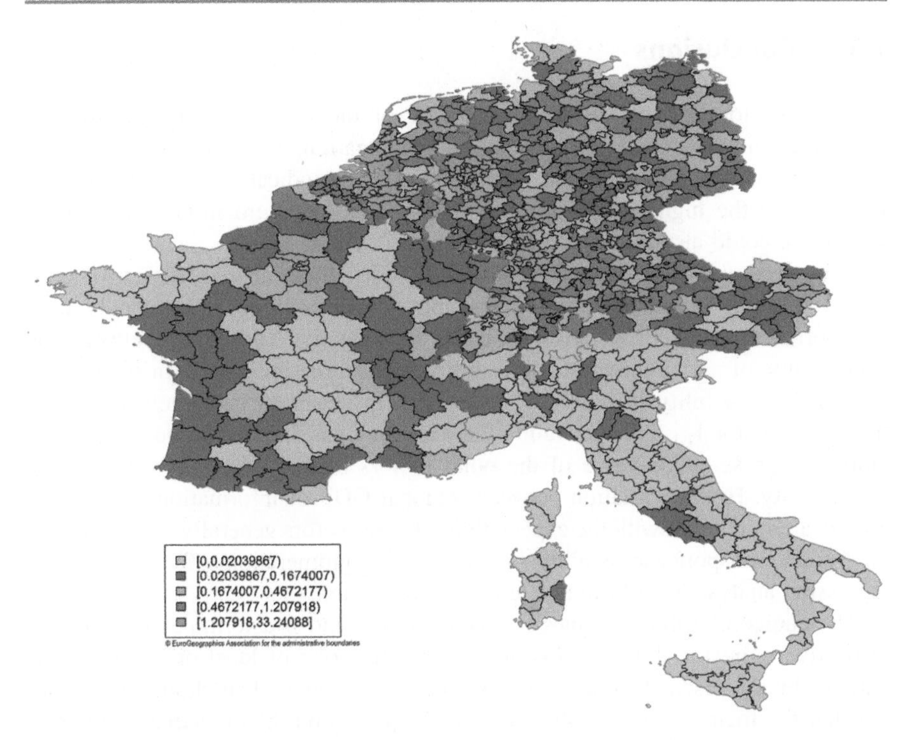


**Fig. 7.4** Regional accessibility of Indicator $I_{2.2}$ (data classification in quintile intervals)

**Table 7.2** Correlation of accessibility indicators with sector specific GDP

| Sector | $I_{1.1}$ | $I_{1.2}$ | $I_{2.1}$ | $I_{2.2}$ | $I_{3.1}$ | $I_{3.2}$ |
|---|---|---|---|---|---|---|
| Agriculture | −0.047 | −0.113 | −0.048 | −0.129 | −0.100 | −0.176 |
| Industry | 0.276 | 0.290 | 0.201 | 0.279 | 0.181 | 0.123 |
| Construction | 0.441 | 0.357 | 0.386 | 0.340 | 0.120 | −0.047 |
| Trade | 0.597 | 0.565 | 0.422 | 0.462 | 0.157 | 0.037 |
| Information[a] | 0.727 | 0.715 | 0.413 | 0.520 | 0.122 | 0.057 |
| Finance[a] | 0.503 | 0.568 | 0.269 | 0.388 | 0.135 | 0.100 |
| Services[a] | 0.508 | 0.536 | 0.418 | 0.487 | 0.076 | 0.003 |
| Other services[b] | 0.560 | 0.543 | 0.377 | 0.435 | 0.154 | 0.050 |
| Manufacturing | 0.125 | 0.182 | 0.143 | 0.252 | 0.087 | 0.130 |

[a]Not available for Germany
[b]Collective variable for services in information, finance, etc

## 7.8 Conclusions

These indicators reflect the actual structure of the transport network from an economic point of view. Depending on the specification, the indicator values stand for rather short or longer transport destinations. The indicators $I_{1.1}$ and $I_{2.2}$ show very clearly the higher regional accessibility of the Rotterdam-Genoa corridor, which one could also assume.

The correlation analysis shows only weak correlation characteristics for indicators representing all transport modes but high correlations when only road transport is considered. Relatively clear is the negative correlation between the added value of agriculture and accessibility in all three indicators. In rural areas, potential accessibility is generally lower than in urban areas. In relation to the theory, indicator $I_{1.1}$–$I_{2.2}$ best fulfils the expectations. They show positive correlations of gross added value of the other sectors of the economy with regional accessibility. High correlation between regional GDP of information and finance sectors are in conflict with the expectations. These sectors generally do not depend on freight transport accessibility but more on agglomeration effects. A further regression analysis could illuminate this contradiction.

The choice of suitable values for parameters $\alpha$ and $\beta$ determines whether the respective indicator shows accessibility rather in terms of local or remote destinations. Future research should already take into account when choosing a model whether the focus of the investigation will be placed on the immediate economic environment or on the accessibility of distant economic zones. It is also important if a consideration of agglomeration effects and the consequent greater emphasis on metropolitan areas is desired—even this question can be included in the analysis by means of an appropriate choice of parameters.

A number of potential future issues come to light against the background of this article. For example, it would make sense to do case studies that relate to urban areas of several NUTS 3 regions. A more detailed consideration of such regions could provide information about accessibility barriers that are only poorly portrayed at the regional level. A classification of such case studies is being developed in the Code24 project.

## References

1. European Commission (2011) Roadmap to a single European transport area—towards a competitive and resource-efficient transport system. White paper on transport, Brussels
2. European Commission (2013) EU Cohesion policy contributing to employment and growth in Europe. Joint paper from the directorates-general for regional and urban policy and employment. Social Affairs and Inclusion, Brussels
3. Bundesministerium für Verkehr und digitale Mobilität (2013) Grundkonzeption für den Bundesverkehrswegeplan 2015. Bedarfsgerecht-transparent-herausfordernd. Report, Berlin
4. Abele G (2009) Transportwirtschaft. Einzelwirtschaftliche und Gesamtwirtschaftliche Grundlagen. Oldenbourg, München
5. Samuelson P (1948) Economics: an introductory analysis. McGraw-Hill, New York

6. Deutsche Bank Research (2003) Road operation projects: lucrative for institutional investors. Frankfurt voice, Frankfurt, 10 June
7. Press release Meldung "Bundesmittel stehen fest". In: Eisenbahn-Revue international 5/ 2004, p 194
8. Bundesverband der deutschen Industrie et al (2012) Damit Deutschland wirtschaftlich stark bleibt. In die Verkehrsinfrastruktur investieren, die Grundlage des Wachstums sichern. Common Declaration, Berlin
9. Scholl B, Braun C, Ghünter F, Otsuka N, Tosoni I (2013) Code24—a common strategy for the corridor Rotterdam-Genoa. Report WP1, Zurich
10. Litman T (2010) Evaluating transportation economic development impacts. Victoria Transport Policy Institute, Victoria
11. Krugman P, Obstfeld M, Melitz M (2012) International economics. Pearson, Essex
12. Hummels D (2007) Transportation costs and international trade in the second era of globalization. J Econ Perspectives 21(3):131–154
13. Harrigan J, Venables A (2006) Timeliness and agglomeration. J Urban Econ 59(2):300–316
14. Nordås H, Pinali E, Geloso Grosso M (2006) Logistics and time as a trade barrier, OECD Trade Directorate. Working papers 35, Paris
15. Hummels D, Skiba A (2004) Shipping the good apples out: an empirical confirmation of the Alchian-Allen conjecture. J Polit Econ 112(6):1384–1402
16. Weber A (1909) Über den Standort der Industrie: Reine Theorie des Standorts. Mohr, Tübingen
17. Christaller W (1933) Die zentralen Orte in Süddeutschland. Gustav Fischer Verlag, Jena
18. Lösch A (1944) Die räumliche Ordnung der Wirtschaft, 2nd edn. Gustav Fischer Verlag, Jena
19. Biehl D (1991) The role of infrastructure in regional development. In: Vickerman RW (ed) Infrastructure and regional development. European research in regional science, vol 1. Pion Ltd, London, p 9–35
20. Blum U (1982) Effects of transportation investments on regional growth: a theoretical and empirical investigation. Pap Reg Sci Assoc 49:169–184
21. Wegener M, Bökemann D (1998) The SASI model: model structure. Deliverable D8 of the EU project socio-economic and spatial impacts of transport infrastructure investments and transport system improvements (SASI), Dortmund
22. Holzhey W (2010) Rail network 2025/2030, Expansion concept for an efficient rail freight service in Germany. Report for Umweltbundesamt (ed), Dessau-Rosslau
23. Rothengatter W (1996) Bottlenecks in European transport infrastructure. Paper presented on the European transport conference, Brunel University, Uxbridge, 01 Jan 1997
24. Cipolina S, Ghiara H (2011) Market situation and context analysis. Project report MoS24, Genoa
25. Drewello H, Günther F (2012) Bottlenecks in railway infrastructure—do they really exist? The Corridor Rotterdam-Genoa. Paper presented on the European transport conference, Crowne Plaza Hotel, Glasgow, 10 Aug 2012
26. Spiekermann K, Wegener M (2006) Accessibility and spatial development in Europe. Scienze Regionali 5(2):15–46
27. Stewart JQ (1947) Empirical mathematical rules concerning the distribution and equilibrium of population. Geogr Rev 37:461–485
28. Wegener M, Eskelinen H, Fürst F, Schürmann C, Spiekermann K (2001) Kriterien der räumlichen Differenzierung des EU-Territoriums: Geografische Lage—Studienprogramm zur Europäischen Raumplanung des Bundesministeriums für Verkehr-, Bau- und Wohnungswesen, Bonn
29. Meyer-Rühle O, Beige S, Bozuwa J, Burg R, Erhardt T, Harmsen J, Hua-Kellermann N, Greinus A, Kille C, Kok R, Röhling W, Roth M (2008) Statistical coverage and economic analysis of the logistics sector in the EU (SEALS). Final report prepared for the European Commission, Basel

30. Banister D, Berechman J (2000) Transport investment and economic development. University College London Press, London
31. CIA World Factbook (2013) Released at https://www.cia.gov/library/publications/the-world-factbook/, on 13/12/16
32. IMF World Economic Outlook Database, released at http://www.imf.org/external/pubs/ft/weo/2013/02/weodata/index.aspx, on 13/12/16
33. PTV (2009) VISUM—State-of-the-art travel demand modeling. Report Karlsruhe
34. Eichhorn C (2011) CODE 24 Transport odel—concept. Code24 report, Karlsruhe May 2011
35. Bleisch A (2005) Die Erreichbarkeit von Regionen: Ein Benchmarking-Modell, dissertation, Basel, released at http://edoc.unibas.ch/277/1/DissB_7206.pdf on 13/12/10
36. Drewello H, Weiß F (2014) Correlation analysis between regional accessibility and sector output for regions along the transport corridor Rotterdam-Genoa. 4th Code24 report of action 9, Kehl Feb 2014

# Part III
# Theoretical Contribution: Assessment Models and Evaluation Frameworks

# Discrete Choice Analysis

# 8

Gerard de Jong and Eric Kroes

**Abstract**

This chapter gives an overview of discrete choice analysis techniques. First we present a reflection about the meaning of the words 'discrete' and 'choice'. Then we provide an overview of the sorts of choices in passenger and freight transport that have been treated as discrete choice problems. The next section presents the basic random utility theory, upon which most discrete choice models have been based. Different types of discrete choice models are then discussed: the workhorse of discrete choice modelling—the multinomial logit model (MNL), the nested logit and other Generalised Extreme Value (GEV) models, the probit model, the mixed logit and latent class models, ordered response models and aggregate logit models. Then we briefly discuss the estimation of discrete choice models and their application for demand forecasting and for policy simulation. The last section contains a summary and conclusions and a discussion on the future research directions.

G. de Jong (✉)
Significance and ITS University of Leeds, Koninginnegracht 23, 2514 AB,
The Hague, The Netherlands
e-mail: dejong@significance.nl

E. Kroes
Significance and VU University of Amsterdam, Koninginnegracht 23, 2514 AB,
The Hague, The Netherlands
e-mail: kroes@significance.nl

I. M. Lami (ed.), *Analytical Decision-Making Methods for Evaluating Sustainable Transport in European Corridors*, Sxi 11, DOI: 10.1007/978-3-319-04786-7_8,

## 8.1 Introduction

### 8.1.1 A Short Reflection on the Word 'Discrete'

In many fields of science the dominant econometric technique is ordinary least squares regression analysis and the many extensions of this framework. In transport analysis, discrete choice analysis plays a relatively important role. The main reason for this is that many of the phenomena that we want to explain (the endogenous or dependent variables) in transport are discrete choices. Well-known examples are mode choice, destination choice or vehicle type choice. In all these examples, the endogenous variable has a purely qualitative character, e.g. representing the modes car, bus and train, or available destination zones. It can only take a limited number of values and whether we label these 0, 1, 2, ..., n or 100, 99, ..., m does not really matter, as long as we can enumerate the set of available alternatives, the choice set. The opposite of a discrete variable, is a continuous variable, which can take any value (possibly within a certain range). A type of dependent variable that is also included in discrete choice (but not in qualitative choice) is an ordered variable. Such variables can also only take a limited number of values, but these values are not just labels, but also determine the ranking. An example is the number of cars per household, where the values 0, 1, 2 and 3 are not just labels representing say different modes but where 3 is more than 2, which is more than 1, etc. Another example would be an answer in a survey on a five-point scale (very bad, bad, neutral, good, very good). For explaining such ordered variables, there are specific ordered response models.

### 8.1.2 A Short Reflection on the Word 'Choice'

The word 'choice' implies that there is some decision-maker who selects one alternative from a number of available options. Indeed, most discrete choice models are about such situations. These models are by their nature also disaggregate models. By 'disaggregate' we mean here that the unit of observation is the individual decision-maker (in passenger transport these are the travellers, in freight transport these are the firms). In 'aggregate' models on the other hand, the units of observation are aggregates of decision-makers, usually in the form of geographical zones. Disaggregate models are much more common in passenger transport than in freight transport. The main reason for this difference is the lack of publicly available disaggregate data on freight transport, which in turn is too a large extent due to the commercial nature of such data. Firms involved in freight transport are often reluctant to disclose information on individual shipments, mode chosen, transport cost, etc. For travellers there often is information on their travel choices from trip diaries where they record their daily (or for more days) trip-making. This is revealed preference information, meaning data on choices made in practice. Both for passenger and freight transport, choice information can also come from stated preferences surveys where travellers, shippers and carriers are asked to choose between hypothetical alternatives in an experimental situation designed by the researcher. For these choice problems by individual decision-makers, there also is a theoretical (micro-economic) foundation for a particular class of discrete choice models, called the 'random utility model' (RUM).

However, not all discrete choice models are disaggregate models. Especially in freight transport (for the reasons stated above) the aggregate discrete choice model is the most commonly used model for mode choice. And this model has also been used in passenger transport. In these aggregate choice models, the data are specified at the level of zones, and the dependent variable in practice usually is the modal share within the transport flow between two zones. We think it is best to see this as a pragmatic approach, not as a model based indirectly on a theory of individual behaviour. However, it regularly leads to satisfactory results (elasticity, forecasts) at a relatively low effort (especially in data collection), both in passenger and freight transport.

### 8.1.3 Overview of this Chapter

In Sect. 8.2 of this chapter, we will provide an overview of choices in passenger and freight transport that have been treated as discrete choice problems. In Sect. 8.3, the basic random utility theory is presented, as well as a short discussion on extensions and alternatives for it. Different types of discrete choice models are discussed in the following sections: the workhorse of discrete choice modelling, the multinomial logit model (MNL), in Sect. 8.4, the nested logit and other Generalised Extreme Value (GEV) models in Sect. 8.5, the probit model in Sect. 8.6, the mixed logit and latent class models in Sect. 8.7, ordered response models in Sect. 8.8 and aggregate logit models in Sect. 8.9. In Sect. 8.10 we will briefly discuss the estimation of discrete choice models and in Sect. 8.11 their application (for forecasting and policy simulation). Finally in Sect. 8.12 we will provide a summary and conclusions and a discussion on the future research directions (e.g. dynamic models, interactions between decision-makers, more integration of micro-economics and behavioural psychology). In this chapter, we can only give some flavour of discrete choice analysis. For a more detailed exposition, we referto the excellent textbooks in this field such as Ben-Akiva and Lerman [2] and Train [36].

## 8.2 Examples of Discrete Choices in Passenger and Freight Transport

Many of the decision problems that are regularly modelled in transport analysis are discrete[1]. Examples of such choices in passenger transport are:

[1] There are also some examples of discrete-continuous choices in transport (e.g. Train [35], de Jong [13], Bhat [4]) and of duration modeling (see de Jong [14], Bhat [5]).

- Mode choice (e.g. choice between car-driver, car-passenger, train, bus/tram/metro, bicycle and walking).
- Destination choice (choice between a number of available destination zones for instance for a shopping trip).
- Tour frequency choice (e.g. choice on the number of trips for a certain purpose on a day).
- Route choice (e.g. between competing car routes in a small road network).
- Departure time choice (choice between a number of time periods for leaving the home, such as morning peak and before the morning peak).
- Car ownership (choice on the number of cars in the household).
- Vehicle type choice (choice between alternatives such as a new small petrol car or a 1–3 years old medium-size diesel car).

Most of these choice situations (mode choice, route choice, departure time choice) can also occur in freight transport. More specific for freight transport are (also see [34]):

- Transport chain choice (the choice alternatives are a consecutive series of modes used between a sender and a receiver of the goods, such as truck–train–truck, or small truck–large truck–small truck; in between these modes, there can be consolidation, distribution and storage locations).
- Supplier choice (in practice this is the choice of the origin zone of the goods by the receiver of the goods; the reverse of destination choice modeling in passenger transport).

Transport chain choice could also be used in passenger transport, since there can also be mode chains for travellers (e.g. walk–bus–train–walk), but this is more often covered by specific choice models for the main mode and for access and egress modes.

## 8.3 The Random Utility Model and Some Alternatives

### 8.3.1 Basic Random Utility Theory

We assume that there is a decision-maker k: a traveller (possibly a group of people travelling together) in passenger transport, a shipper, carrier, receiver or truck driver in freight transport, who decides on a choice of selecting an alternative i from a number of discrete alternatives I. Furthermore, we assume that the decision-maker decides by comparing the different alternatives available to him or her and choosing the one that gives him or her the highest utility: the most widely used choice paradigm and theoretical foundation is that of random utility maximisation, RUM [24–26]. The basic equation of the RUM model is:

$$U_{ik} = G_{ik} + e_{ik} \tag{8.1}$$

in which:

$U_{ik}$: Utility that decision-maker k derives from choice alternative i (k = 1,…K; i = 1,…,I).
$G_{ik}$: Observed (or 'structural') utility component.
$\varepsilon_{ik}$: Unobserved or random utility component.
Utility maximisation comes from the economics of consumer behaviour.

As a more specific example, for the choice of mode for a specific trip by decision-maker k, between three alternatives (car, train and bus), one might specify the following linear[2] utility functions[3]:

$$U_{car} = \beta_0 + \beta_1 \cdot COST_{car} + \beta_2 \cdot TIME_{car} + \beta_3 \cdot REL_{car} + e_{car}, \quad (8.2a)$$

$$U_{train} = \beta_4 + \beta_1 \cdot COST_{train} + \beta_5 \cdot TIME_{train} + \beta_6 \cdot REL_{train} + e_{train}, \quad (8.2b)$$

$$U_{bus} = \beta_1 \cdot COST_{bus} + \beta_7 \cdot TIME_{bus} + \beta_8 \cdot REL_{bus} + e_{bus}, \quad (8.2c)$$

in which:

$COST_i$: Travel cost of mode i.
$TIME_i$: Travel time of mode i in hours.
$REL_i$: Journey time reliability of mode i; this could be measured as the standard deviation of travel time.
$\beta_0$, $\beta_1$, …,$\beta_8$: Coefficients to be estimated; we expect negative signs for $\beta_1$,… $\beta_3$ and $\beta_5$, … $\beta_8$, the sign for $\beta_0$ and $\beta_4$ can be positive or negative.

In Eqs. (8.2a)–(8.2c), the utility that would be obtained when choosing the car depends on the travel cost, time and travel time reliability for that trip by car, and likewise for the other two modes. The values for COST, TIME and REL by mode, may come from skimming networks for these modes, but also might be provided by the travellers themselves (however, often they find this hard for non-chosen modes, and there could be perception errors in their answers) or have been postulated in a 'what if' fashion by the researcher in a stated preference survey.

The $\beta$s are coefficients for which numerical values are determined by estimating the model on data for various decision-makers and their corresponding trips (which may vary in terms of origins and destinations, leading to variation in distance and time by mode between trips). This could for instance be a trip diary survey. $\beta_0$ and $\beta_4$ are so-called 'alternative-specific constants', ASCs. There can

[2] Non-linear specifications of the utility function, such as functions with logarithmic or Box-Cox transformations (e.g. [28]), are also possible.

[3] Strictly speaking there is also a 'scale' parameter, which reflects the variance of the random component of utility and is used for normalising the model. It is called 'scale' parameter, because it scales the $\beta$ parameters in (8.2a)–(8.2c); a higher random variance leads to lower estimated values of the $\beta$' s.

only be N-1 ASCs in a discrete choice model, N being the number of available choice alternatives, because in a utility maximisation model only differences in observed utility matter. In the example above, bus is the alternative without an ASC, which means that for this alternative, the constant is normalised to 0. Attributes can only be included as explanatory variables if they differ between alternatives, again because only differences in observed utility matter. Attributes of the decision-maker (e.g. the age of the person) or of the trip (e.g. in the peak or not) can only be included by interacting these variables with characteristics of the modes. An example of such an interaction would be making certain travellers less cost-sensitive or including a school age dummy only for bus, expressing that school age persons are more likely to be travel by bus.

Coefficients can be generic, the same for all alternatives, such as $\beta_1$ for cost above, or alternative-specific, such as the other coefficients in Eqs. (8.2a)–(8.2c). Which specification is best is largely an empirical matter, so various forms should be tested and compared against each other. In the above model specification we have used generic coefficients for costs, but not for the other variables. An advantage of this specification is that one unit of money paid for road transport has the same value as one unit of money paid for train or bus ('a euro is a euro').

Decision-maker k's mode choice decision could be influenced by other variables than the three attributes included above. For instance the flexibility of a mode, the service frequency and the level of comfort provided by bus and train might also play a role. But the researcher that is constructing the mode choice model may not have any data on these influencing variables, or only data measured with some error. This is a key reason[4] for including the error components $e_{car}$, $e_{train}$ and $e_{bus}$ to the utility functions: they represent variables that affect the utility of the decision-maker, but that are not observed by the researcher (or that are only observed with some measurement errors).

Now the researcher, in order to deal with the error components, assumes these are random variables (with a mean of zero and some variance), and by making different assumptions on the probability distribution of the error components, different discrete choice models can be derived. These models are also called 'probability models', because they do not generate a certain choice, but probabilities for each of the available alternatives.

### 8.3.2 Non-RUM Models

As discussed above, discrete choice models can be interpreted as random utility maximisation models, and in this way these models are founded in micro-economic theory and represent 'rational' behaviour. However, this economic perspective is just one way of looking at decision-making by humans or firms. There are other perspectives on behaviour (e.g. as studied by psychology or sociology) as well. For some decisions the economic perspective might be more relevant (here

[4] There are other reasons in the discrete choice literature for including the error terms.

one might think of important well-planned long-run decisions, especially in a business context), for other decisions other perspectives might be more important. Until recently, RUM was the only available serious candidate as a foundation for mathematical models of making discrete choices. In recent years two other paradigms have gained some popularity in transport modelling, namely prospect theory and regret minimisation.

Prospect theory ([21, 22]; for an extended form also see [37]) was developed by psychologists, who found that in many experimental situations, respondents did not follow the theory of rational behaviour. They then tried to stay closer to how people actually respond in these experiments. Insights from prospect theory include the following notions:

- The valuation of an attribute depends on the current value of that attribute, i.e. it depends on the reference alternative (the situation as observed now): this is called 'reference dependence'.
- There is a difference in the valuation of gains and losses in an attribute: losses loom larger, per unit: this is called 'loss aversion'.
- There is a difference in the valuation at different values of an attribute (e.g. between a short and a long transport): this is called 'size dependence'.

Instead of basing discrete choice models on utility maximisation, they can also be based on the minimisation of anticipated regret by the decision-makers. Regret occurs when a non-chosen alternative scores better on some attribute than the chosen alternative. The random regret model then distinguishes two regret components in the regret function, 'observed' or 'systematic' regret and 'random' regret, leading to the random regret model RRM [8]. Systematic regret is calculated by doing a pairwise comparison of available alternatives on the basis of the attribute values of these two alternatives, and then summing the outcome (the binary regret, defined as the sum of the attribute differences between the two alternatives) over all pairs.

The distributional assumptions for the random component(s) can be similar to those of the RUM model, and sometimes even the same software can be used for RRM and RUM, but the choice probabilities from the two types of models will be different. The key difference here is that RRM captures the 'compromise effect': alternatives that perform 'in-between' on all attributes, relative to the other alternatives, are generally favoured over alternatives with a poor performance on some attributes and a strong performance on others [9].

## 8.4 The Multinomial Logit Model

Different distributional assumptions lead to different discrete choice models. The most common assumption for the error components e, both in passenger and freight modelling, is that they are independently and identically (i.e. same variance across observations) distributed following the extreme value distribution type I (or Gumbel distribution). This leads to the multinomial logit[5] (MNL) model with the choice probabilities:

$$P_{ik} = \frac{e^{G_{ik}}}{\sum_i e^{G_{ik}}}. \tag{8.3}$$

A well-known restriction of the MNL model is that the cross-elasticities are the same: if in the mode choice model in Eqs. (8.2a)–(8.2c) the cost of road transport increases, substitution will occur to rail and inland waterways in proportion to their current market shares, so that the road cost elasticities of demand for rail transport and for inland waterway transport will be the same. Another manifestation of basically the same phenomenon (which is due to the independence of the error terms) is the independence from irrelevant alternatives (IIA) property: the ratio of the choice probabilities between two alternatives does not depend on any other alternative. These properties may be at odds with reality. In practice, there could for instance be more substitution between rail and inland waterways than between any of these alternatives and road transport.

## 8.5 The Nested Logit and Other GEV Models

A relatively easy way to accommodate for this is the nested logit (NL) model (e.g. [12]) in which rail and inland waterways would be grouped in a nest, allowing correlation between these alternatives (Fig. 8.1).

Mathematically, the easiest representation is to distinguish two probabilities (as for instance in [36]), linked to each other by the logsum variable.

$$P_{B_l k} = \frac{e^{G_{lk} + \lambda_l I_{lk}}}{\sum_l e^{G_{lk} + \lambda_l I_{lk}}}, \tag{8.4a}$$

$$P_{\langle ik|B_l \rangle} = \frac{e^{H_{ik}/\lambda_l}}{\sum_{i \in B_l} e^{H_{ik}/\lambda_l}}, \tag{8.4b}$$

$$I_{lk} = \ln \sum_{i \in B_l} e^{H_{ik}/\lambda_l}. \tag{8.4c}$$

[5] The name 'logit' comes from the fact that in the model presented above we are actually looking at differences between utility functions and the difference of two Gumbel distributions follows a logistic distribution. A 'logistic' or 'logit' model, that was invoked by assuming a logistic distribution for the error terms, had been around long before the above RUM-based model was first presented by especially McFadden (see [10]).

**Fig. 8.1** Nested logit structure for freight mode choice

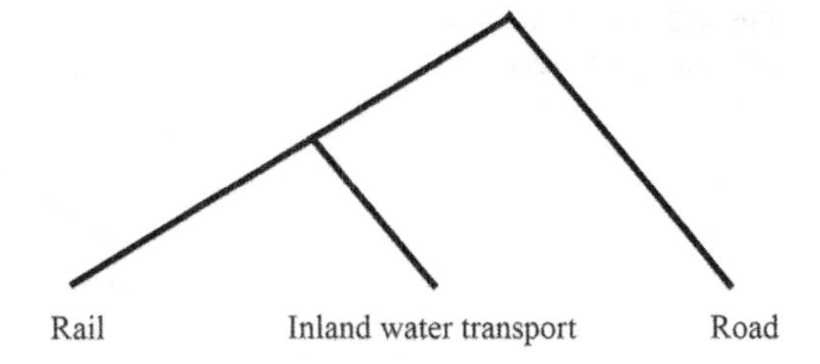


The first probability (8.4a) gives the chance that decision-maker k chooses an alternative within nest $B_l$. This depends on the generalised cost G of l plus a coefficient $\lambda_l$ times the expected cost from the alternatives in the nest, represented by $I_{lk}$, the so-called 'logsum' variable, relative to the same kind of costs for the all alternatives at the nesting level.[6]

The second (conditional) probability gives the change of choosing alternative i given that nest $B_l$ has been chosen. This depends on the generalised cost of this alternative relative to those for all alternatives in the nest.

Now the unconditional probability that decision-maker k will select alternative i is:

$$P_{ik} = P_{\langle ik|B_l \rangle} P_{B_l k}. \tag{8.4d}$$

The coefficient $\lambda_l$ is the 'logsum coefficient' which gives the degree of correlation between the error components of the alternatives in nest $B_l$: the higher this coefficient, the lower the correlation. In estimation, this is an extra parameter to be estimated. The estimated value must be between 0 and 1 for global consistency (meaning: across the entire range for the exogenous variables) with RUM. If a value above 1 is found, this is an indication that a different (especially a reversed) nesting structure would work better and be consistent with RUM.

Both MNL and nested logit are members of a family of models, the GEV family [11, 25] which contain more members (and still sometimes new members are discovered).All of these are consistent with random utility maximisation. Most of these have only seen a limited number of applications in passenger transport and none or almost none in freight transport, though they offer more flexibility in terms of substitution patterns between alternatives than MNL or nested logit. One such member is the cross-nested logit [1, 39] where a single alternative can belong to more than one nest at the same time. An example is given in Fig. 8.2, where rail is in a nest with inland waterway transport (as before), but also in another nest with the other non-water-based mode, road transport. In this structure, inland waterway

---

[6] The same logsum variable also has a meaning in welfare analysis: this logsum can be converted into money units (e.g. using a costs coefficient) and then the difference between the monetised logsum for a project alternative (with some infrastructure project) and the one for a reference case (without the project) can be used as the change in consumer surplus as a result of the project [15].

**Fig. 8.2** Cross-nested logit structure for freight mode choice

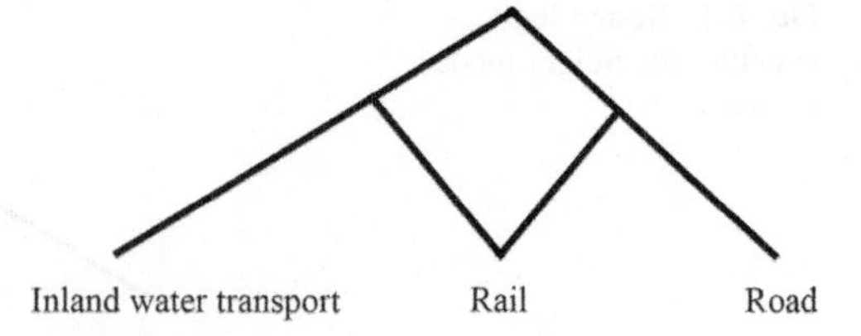


transport and road transport are not correlated with each other, but rail is correlated with both.

The corresponding equations for CNL are:

$$P_{B_l k} = \frac{e^{G_{lk} + \lambda_l I_{lk}}}{\sum_l e^{G_{lk} + \lambda_l I_{lk}}}, \tag{8.5a}$$

$$P_{\langle ik|B_l \rangle} = \frac{\alpha_{il} e^{H_{ik}/\lambda_l}}{\sum_{i \in B_l} \alpha_{il} e^{H_{ik}/\lambda_l}}, \tag{8.5b}$$

$$I_{lk} = \ln \sum_{i \in B_l} \alpha_{il} e^{H_{ik}/\lambda_l}, \tag{8.5c}$$

$$P_{ik} = \sum_l P_{\langle ik|B_l \rangle} \cdot P_{B_l k}. \tag{8.5d}$$

In Eqs. (8.5a)–(8.5d), l stands for a specific nesting structure for alternative i (for instance 'slow modes nesting' and 'land-modes nesting'). There now is an extra parameter $\alpha_{il}$: the share of the l'th nesting structure for alternative i (the degree of membership of the nest). Over all nesting structures l, the $\alpha_{il}$ should sum to 1. When one would have two nesting structures (as in the example above), the standard pragmatic procedure is to give both a weight of ½. Hess et al. [18] also used ½, which they motivated saying that the estimation of these membership coefficients $\alpha_{il}$ would have been very difficult.

Yet another GEV model, is the ordered extreme value (OGEV) model [33]. This model presupposes ordered[7] alternatives (e.g. time periods during one day), whereas mode choice is unordered. In the OGEV model there is more substitution between an alternative and its direct neighbours (e.g. adjacent time periods) than between the alternative and alternatives further away from it in the ordering (e.g. time periods that are separated from each other by other periods).

---

[7] Yet this is not the same model as the ordered response model, that is described in Sect. 8.8 of this chapter. OGEV is a generalisation of MNL, whereas ordered response models assume the same (non-RUM) response mechanism behind all ordered categories.

## 8.6 The Probit Model

Assuming that the error components follow a multivariate Normal distribution leads to the probit model. This model is almost as straightforward to estimate as MNL when there are only two alternatives.

Probit models with more than two alternatives lead to multidimensional integrals, and simulation methods then have to be used for estimation (e.g. [16]).This is all (still) relatively cumbersome. On the other hand, the probit model is very flexible, since the modeller can define and estimate specific correlation parameters between the alternatives. This is mainly relevant if the modeller already has a priori ideas which variables will be correlated with each other and which variables will not.

## 8.7 The Mixed Logit and the Latent Class Model

The mixed logit model (or mixed multinomial logit MMNL model) was developed in the late nineties and saw its breakthrough in the first decade of the 21st century. It has two error components $e_{ik} + v_{ik}$, where the first one follows the Gumbel distribution (so that conditional on the other, the model is MNL) and the other can follow any distribution, specified by the analyst, such as the Normal or lognormal. Mixed logit can accommodate two ideas, which are both generalisations of MNL:

1. It can capture taste variation between decision-makers (random coefficients): Not every respondent has the same coefficient, but the coefficients, such as $\beta_1$ above, follow a statistical distribution of which the mean and the variance are estimated. It is possible to include interaction coefficients for observed heterogeneity and randomness for unobserved heterogeneity in coefficients like those for transport time and cost at the same time. It is also possible in mixed logit to estimate one of several random coefficients (e.g. for transport cost and time) where in taking the random draws one takes account of the fact that draws for the same individual should not be independent. This is important for panel data or for Stated preference experiments where the respondents are asked to make several choices. With such data sets, the researcher cannot assume that all observations are independent, because the observations obtained from the same respondent contain the same (or a similar) error.
2. It can distinguish error components that capture a specific correlation structure between alternatives. In its most general form, the utility function now becomes:

$$U_{ik} = \sum_{r} \beta_r x_{rik} + \sum_{s} \sum_{t} \eta_s w^i_{st} v_t + e_{ik} \tag{8.6}$$

in which:

The first term on the right-hand side represents the influence of the observed attributes on utility.

The second term on the right-hand side determines the error component structure:

- $v_t$ is an error component, following some statistical distribution f(0, 1), which can consist of several random subcomponents (t = 1, …,T).
- $\eta_s$ is a coefficient to be estimated.
- $w^i_{st}$ is a general weighting matrix, based on data and/or fixed by the researcher, for alternative i, with rows s corresponding to the coefficients $\eta$ and columns t corresponding to the error subcomponents in v.

The third term on the right-hand side $e_{ik}$ is the error component. In the mixed logit model it is Gumbel distributed (if $v_t$ and $e_{ik}$ would follow the multivariate Normal distribution, the model would be multinomial probit).

Many models in the scientific journals in transport, especially in passenger transport, nowadays use mixed logit. Often the estimation data are stated preference data and the purpose of the estimation is to provide monetary values for non-monetary attributes, such as transport time (value of travel time), reliability (value of travel time variability). The SP alternatives in these experiments and models are usually 'abstract' or 'unlabelled' alternatives, meaning that it is not specified whether the choice alternatives are different modes, route, carriers, etc. Such studies only require that the mixed logit simulation of the non-Gumbel distribution (e.g. drawing many times from a lognormal distribution) is done for each model specification that is estimated. For transport models that are developed for use in forecasting for scenarios and for the impact of transport projects and policies relative to a reference scenario, mixed logit is still not very attractive because every time the model is applied, the random draws need to be made again, which is very time consuming. In the framework of a large model system (e.g. with many zone pairs), possibly with feedback loops, this would simply take too long. Hardly any practical transport forecasting model, in passenger and freight transport, therefore uses mixed logit.[8]

A disadvantage of mixed logit with continuous statistical distributions for both error components is that the researcher has to assume the shape of the distributions (e.g. normal or lognormal distribution) and this assumption can strongly affect the final estimation and application results. In addition, this technique does not always yield a stable estimation result. A better choice could then the latent class model, which is related to mixed logit. Instead of using continuous distributions of coefficients, as in the mixed logit random coefficients specification, this model uses a discrete number of possible values (latent classes) for a coefficient like the cost coefficient. Furthermore, membership equations can be estimated, linking membership to one of the classes to observed variables (which can include specially collected attitudinal variables, but this is not necessary). In Hess et al. [19], latent

[8] An exception, in passenger transport, is the SILVESTER model for Stockholm [23].

class models were compared against mixed logit models with a continuous distribution, with a positive outcome for latent class models. Similar conclusions on latent class models were reached in Greene and Hensher [17].

## 8.8 Ordered Response Models

The ordered response model is rather different from the discrete choice models discussed above, and some authors do not include this model among the discrete choice models. It does not share the foundation in random utility maximization. On the other hand, this model can handle the ordered nature of a dependent variable, such as a response ranging from 'strongly disagree' to 'strongly agree', income bands, the number of trips on a survey day or the number of cars in a household: any variable where the different outcomes form a specific order (but not necessarily have a quantitative meaning; though it is sometimes used if the dependent variable is quantitative but can only take a few values[9]).

The ordered regression model postulates that there is an underlying (latent) dependent variable y*, that is continuous, but that the researcher cannot observe. This variable is a function of observed exogenous variables x and some error term $\varepsilon$:

$$y^* = X'\beta + \varepsilon \tag{8.7}$$

where:

y*: underlying (unobserved) continuous dependent variable.
X: vector of explanatory variables.
$\beta$: a vector of coefficients to be estimated.
$\varepsilon$: the error term.

What the researcher can observe is the category of the dependent variable for every observation:

$$\begin{array}{ll} y = 0 & \text{if } y* \leq \mu_1 \\ y = 1 & \text{if } \mu_1 < y* \leq \mu_2 \\ y = 2 & \text{if } \mu_2 < y* \leq \mu_3 \\ y = N & \text{if } \mu_N < y* \end{array} \tag{8.8}$$

in which:

y: the ordered (observed) dependent variable.
$\mu_n$: threshold coefficient to be estimated (there are N of these: the number of categories for y minus 1; since we have N + 1 categories or possible outcomes for y).

[9] For the latter problem an alternative category of models is formed by the count data models, such as the Poisson regression model (see [7]).

If one assumes that the error term follows the logistic distribution, then the model becomes ordered logit. An alternative is assuming a Normal distribution which leads to ordered probit.

A somewhat restrictive feature of the ordered response model is that the same mechanism and $\beta$ coefficients and explanatory variables are used for all categories. The ordered response model also cannot handle alternative-specific variables, but only attributes of the decision-maker (in X). For this reason a multinomial logit model, which allows for different utility functions for different alternatives may in some cases lead to better predictions of the observed outcomes than the ordered response model, even though it does not capture the ordering [6].

## 8.9 Aggregate Logit Models

As discussed in the introduction of this chapter, the standard disaggregate choice model (MNL), and many extensions of it, can be based on the theory of utility maximisation by individual decision-makers. There also is an aggregate form of this model (often called 'aggregate logit') where the observations usually refer to summations of trips or shipments for the same origin-destination (OD) zone pair. More specifically, these modal split models are estimated on data for the market share of each mode over different OD pairs. The aggregate modal split model can indirectly be based on the theory of individual utility maximisation, where all decision-makers on the trips or shipments for an OD pair are optimising their subjective utility, but only under very restrictive assumptions. These assumptions basically boil down to assuming that all variation in characteristics of the decision-makers and of the trips/goods belongs to the error component of the utility function. This would be such a far-reaching assumption, that it is better to see aggregate logit models as pragmatic models (that have shown to be able to produce plausible results) instead of models based on a theory of behaviour.

The aggregate logit models are often selected because disaggregate data are not available. This typically happens, especially but not exclusively in freight transport. In freight transport the analyst frequently only has data on the tonnes by OD zone pair and mode (or tonnes by PC pair and main mode). The aggregate logit can easily be estimated (both with software for linear regression models and for discrete choice models), produces an intuitively appealing S-shaped market shares curve and market shares, which are always between 0 and 1. Because of these advantages, the aggregate logit model still is the single model specification used most in practical freight mode choice modelling.[10]

[10] In practice it is often even difficult to obtain plausible transport time and costs coefficients when estimating on aggregate data. Prof. Moshe Ben-Akiva once suggested here to assume a value of time distribution to allow for heterogeneity between shipments.

A typical formulation is the 'difference form'[11]:

$$\log\frac{S_i}{S_j} = \beta_0 + \beta_1(P_i - P_j) + \sum_w \beta_w(x_{iw} - x_{jw}) \tag{8.9}$$

in which:

$S_i/S_j$: is the ratio of the market share of mode i to the market share of mode j.
$P_i$ and $P_j$: are the transport costs using these two modes.
$X_{iw} - X_{jw}$: are w (w = 1,..., W) differences in other characteristics of the two modes.

This model can be estimated using specialised discrete choice estimation software (the same as used for estimating disaggregate models), but also by standard regression analysis (regressing the log-ratio above on its explanatory variables).

Aggregate modal split models are mostly binomial (two available modes) or multinomial logit models (three or more available modes). In case of more than two alternatives, all the different pairwise combinations of alternatives have to be taken into account. Since aggregate model split models only give the market share of a mode, not the absolute amount of transport (tonnes) or traffic (vehicles), the elasticities from such models should be regarded as conditional elasticities (conditional on the quantity demanded; see [3]).

## 8.10 Estimation of Discrete Choice Models

The MNL, NL, CNL and OGEVand latent class models can all be estimated by Maximum Likelihood methods that do not involve any simulation. Several software packages (including some more general ones such as SPSS, Stata and SAS) contain MNL estimation (sometimes called 'conditional logit'), some more specialised packages also NL (e.g. Biogeme, Alogit, Nlogit).But CNL and OGEV are only included in one or two specialised estimation packages (e.g. in Biogeme). Estimation of the mixed logit model requires simulation (see [36]) in the sense of taking a large number of random draws for v and calculating the Likelihood function for each of the draws, but because of the presence of the Gumbel component this can be done much faster than for probit. This facility is included in several estimation packages (Biogeme, Alogit, NLogit). The ordered logit model is included in Biogeme, SPSS, Stata, SAS and Limdep, whereas Alogit has a special facility to estimate aggregate logit models on data on the market shares.

---

[11] An alternative for the difference form is the 'ratio form' where the right-hand side has $P_i/P_j$ and $x_{iw}/x_{jw}$, which has the disadvantage that the choice of the base mode (in the denominator of the dependent variable) affects the estimation results and the elasticities from the model. The difference form does not have this disadvantage.

**Table 8.1** Estimation results for vehicle type choice (standard binary choice logit model)

| Variable | Estimated coefficient | t-value |
|---|---|---|
| Purchase price (euro) | −3.64e−5 | −18.7 |
| Fixed car cost (excluding depreciation; euro) | −1.77e−4 | −4.6 |
| Dummy for diesel car | −0.221 | −3.5 |
| Dummy for car of 10 or more years old | −0.638 | −15.3 |
| Fuel efficiency (km/L) | 0.082 | 11.5 |
| Road user charge per kilometre (eurocent/km) | −0.028 | −2.9 |
| Number of observations | 9,905 | |
| Final log-likelihood value | −5,628.8 | |
| Pseudo-Rho$^2$ w.r.t. 0 | 0.180 | |

*Source* [31]

An example of estimation results is given in Table 8.1. This table gives the estimated coefficients of a standard binary logit model (independent error terms, as in Sect. 8.4) for a choice of individual households between different vehicle types. The data come from a stated preference survey in The Netherlands carried out for the Environmental Assessment Agency [31]. In each choice situation, a household can choose between two vehicle alternatives that are described in terms of their attributes, including purchase price, other fixed costs, fuel type (diesel or petrol), age class, fuel efficiency and type-specific road user charge (as a hypothetical policy measure).

Each respondent evaluated twelve choice situations where he/she had to choose between two different vehicle types, labelled A en B. In the simple model presented above, no correction has been carried out yet for the repeated measurements character of the data (the observations for the same individual will not be independent). This can be dealt with in the mixed logit or latent class models with a panel specification (see Sect. 8.7).

In Table 8.1 we can see from the negative sign of the estimated coefficients that vehicles with a higher purchase price, other fixed cost or kilometre charge have a smaller probability of being chosen. Diesel cars and older cars also lead to a reduction in utility and a smaller chance of being selected (all other things being equal).More fuel efficient cars on the other hand are more attractive. The t-ratios in the table tell us whether an estimated coefficient is significant: a coefficient differs significantly (at the 95 % level) from zero if the t-ratio is less than −1.96 (for negative coefficients) or above 1.96 (for positive coefficients). The final Log-Likelihood value gives the result for the value that was optimised in the estimation (it needs to be as close as possible to 0). It only becomes informative when compared to a benchmark value, such as the Log-Likelihood value for the situation without any coefficients in the model (in other words: when all coefficients are zero). The result of this comparison is expressed here in the form of the Pseudo-Rho-square with respect to zero. This value is between 0 and 1, and the higher the better. It does however not have the same interpretation as the Rho-square in regression (the share of variation that is explained by the regression equation) and the very high Rho-square values that time series regression sometimes finds are much less common for the Pseudo-Rho-square from a Maximum Likelihood estimation on disaggregate data.

In Table 8.2 an estimation result is given for an aggregate logit model (also with independent error terms) for mode choice in freight transport, taken from the estimation for the national freight model of The Netherlands BasGoed v2.0 [32]. This model was estimated on data on the mode shares within each origin-destination flow (for domestic, import and export flows). The modes used are truck, train and inland waterways, though for some OD pairs not all these alternatives are made available. Different models were estimated for different commodity groups; due to space constraints, we only present there the results for the commodity categories NSTR6 (minerals), NSTR7 (fertilisers), NSTR8 (chemicals) and NTSR9 (other goods).

In Table 8.2, we have the estimated coefficients for transport time and cost at the bottom. Time and cost are calculated from mode-specific networks, but here we have specified the same coefficients over the modes (transport cost also includes the time-dependent transport cost; transport time in the model reflects the cargo-related time concerns). If time or cost for some mode becomes longer, this reduces the choice probability of that mode. The Grns_Bvrt and Grns_Spr coefficients are border resistance terms for inland waterway and rail transport, reflecting that the combination of these modes and cross-border transport is less (or more) likely. The coefficients CONT_Bvrt and CONT_Spr relate to variables for the degree of containerisation, interacted with inland waterway or road transport. For instance for other goods (which is the main goods category for container transport), inland waterway transport and rail transport have an increased choice probability relative to road transport when the OD pair has a higher degree of goods in containers. The BasGoed modal split models also contains a large number of zone-specific constants (interacted with two of the three modes), but these are not presented here.

The Pseudo-Rho-square with respect to c compares the final Log-Likelihood value to that of a model with only alternative specific constants (for 3-1 alternatives in this case).

MNL models can be estimated consistently (i.e. without bias) on a sample that is non-representative with regards to the explanatory variables. If the sample is non-representative with regards to the choice variable (e.g. with an overrepresentation of rail transport), and the model has $N - 1$ ASCs and M other coefficients, all M coefficients can still be estimated consistently using standard methods, and only the $N - 1$ ASCs will be biased. These ASCs can be corrected after the estimation by a simple formula based on the observed market shares [26], or estimated by a weighted procedure.

**Table 8.2** Estimation results for mode choice in freight transport, by commodity class (t-ratios are given within brackets)

| Variable/Commodity | NSTR6 | NSTR7 | NSTR8 | NSTR9 |
|---|---|---|---|---|
| Grns_Bvrt | −0.7958 (−36.3) | −0.3886 (−5.4) | 0.5887 (25.8) | −0.4452 (-21.2) |
| Grns_Spr | 0 (*) | 0 (*) | −2.521 (−37.8) | −1.548 (−34.2) |
| CONT_Bvrt | 1.005 (3.7) | 0 (*) | 0 (*) | 6.031 (122.4) |
| CONT_Spr | −154.7 (−12.5) | −60.85 (−5.0) | 2.484 (9.7) | 5.750 (60.8) |
| Time | −4.74e−4 (−38.7) | −6.04e−4 (−13.7) | −3.31e−4 (−38.9) | −3.57e−4 (−60.2) |
| Cost | −0.2732 (-100.9) | −0.2254 (−20.1) | −0.09698 (−54.2) | −0.03366 (−31.3) |
| Number of observations | 1,394 | 450 | 580 | 1,120 |
| Final log-likelihood value | −91,797.6 | −3,284 | −34,833.7 | −61,230.1 |
| Pseudo-$Rho^2$ w.r.t. 0 | 0.364 | 0.453 | 0.382 | 0.575 |
| Pseudo-$Rho^2$ w.r.t. c | 0.339 | 0.44 | 0.278 | 0.441 |

*Source* [32]

## 8.11 Application of Discrete Choice Models

Having estimated the model, one can apply the estimated coefficients on a sample of trips or shipments to calculate probabilities for each choice alternative for each observation. If this sample is representative of the population studied, one can then simply sum the probabilities (this method is called 'sample enumeration') over all observations in the sample to get the population market shares for the alternatives (e.g. the share of car in the total for car, train and bus) as predicted by the model. For non-representative samples, one can do a weighted summation with the population to sample fractions for each observation as weights. Different zones in a study area or different horizon years might even have different sets of weights. Such applications can give the impact of changing a single variable at a time (which can be expressed in the form of elasticities), but can also predict what would happen in case of an input scenario with changes for several (possibly all) variables in the model.

Another way of applying the estimated coefficients is to compute the probabilities for all alternatives and to draw a random number between 0 and 1 to determine which alternative is chosen. This generates a simulated discrete choice per individual. This is called 'micro-simulation' and has been done in several activity-based travel models [27]. In order to produce robust results, this procedure needs to be repeated for a very large number of respondents.

Discrete choice models estimated on stated preference (SP) data only should not directly be used for forecasting (including the derivation of elasticity values), since the SP data will have a different variance for the error component than real world data. This will affect the choice probabilities. It happens because in the experimental set-up of the SP many things that can vary in reality are kept fixed, and vice versa. Therefore, it is better to use revealed preference (RP) data or combined SP/RP data, where the variance of the SP is scaled to that of the RP, for forecasting purposes. Models estimated on SP data alone can directly be used to derive ratios of coefficients. An example of this is the perceived monetary disbenefit of a diesel car (derived by dividing the diesel coefficient by one of the monetary coefficients from Table 8.1). A second example is a value of time, as can be calculated from Table 8.2 by dividing the time by the cost coefficient (please note that this is the value of time per euro per tonne per minute for the goods component only, e.g. capital costs on the value of the goods in transit). A third example would be the value of reliability. In calculating these ratios, the SP error component drops out.

## 8.12 Conclusions

Discrete choice models have been used for analyzing many problems in passenger and freight transport and elsewhere, particularly since the breakthrough in the seventies that provided a theoretical foundation for these models in the theory of random utility maximisation. The specific model that was used most in these early days and still is used most in practice is the multinomial logit model, with the nested logit model ranking second. In academic research, but also in practical research trying to establish monetary values for attributes such as time, reliability and comfort, richer models like the mixed logit and latent class models are used quite frequently nowadays (mostly on Stated Preference data).This is not the case for large-scale forecasting models, which still tend to rely on multinomial and nested logit, mainly to avoid the long run times that repeated simulation from statistical distributions would entail. Other discrete models, such as cross-nested logit, ordered GEV and multinomial probit are only rarely used, partly because of the capability of mixed logit to deal with the issues that these models specialise in (such as flexible correlation structures). Ordered response models are also not often used in transport analysis, even though they are relatively easy to estimate. Their limited use can be explained by the fact that there are not so many ordered response problems in transport and even then, non-ordered models might perform better. Aggregate logit models are still used quite a lot, especially in freight transport mode choice modelling, where disaggregate data often is not available.

In terms of developments that we expect in the coming years, there will probably be a further integration of concepts from micro-economics and behavioural psychology. On the one hand this refers to models that use behavioural decision rules other than random utility theory, such as elements from prospect theory and random regret minimisation. On the other hand, we expect that the trend to include more dependent (possibly latent) and independent variables that measure attitudes, perceptions, intentions and plans will continue, as happens for instance in the latent class models and latent variable models.

Another development that we can already see taking off, and that will probably attract more attention is that of dynamic discrete choice models. All the above models were static (though in mixed logit and latent class models it is possible to take account of the panel nature of the data, such as repeated measurements on the same individuals in SP data). Genuine dynamic models contain dependent variables referring to several periods (time dimension). In a discrete choice setting there can be Markov models that look at changes over time between categories (e.g. mode choice of car ownership). There also is a foundation for dynamic discrete choice models in utility maximisation, looking at current utility and the expectation of future (remaining lifetime) utility [30, 36].

Finally, the above choice models all relate to the behaviour of an individual independent decision-maker; the theory is that each traveller or firm maximises his own subjective utility and is not affected by others (with the exception of a macro-level feedback if congestion occurs, as in several large transport models). But several decisions are in reality the result of interactions between multiple decision-makers. Examples are the mode and shipment size decision in freight transport [20] by shippers and carriers, decision-making on who will use the car(s) in the household by different household members or on the daily activity schedules by household members (e.g. who will do the shopping or bring the children to school?). Discrete choice models for such interactions have already been developed (e.g. [29, 38]) and we are likely to see more of these in the future.

## References

1. Ben-Akiva M, Bierlaire M (1999) Discrete choice methods and their applications to short-term travel decisions. In: Hall R (ed) Handbook of transportation science. Kluwer, Amsterdam
2. Ben-Akiva M, Lerman SR (1985) Discrete choice analysis, theory and application to travel demand. MIT Press, Cambridge, Massachusetts
3. Beuthe M, Jourquin J, Geerts JF, Koul à Ndjang'Ha C (2001) Freight transport demand elasticities: a geographic multimodal transportation network analysis. Transp Res 37:253–266
4. Bhat CR (2008) The multiple discrete-continuous extreme value (MDCEV) model: Role of utility function parameters, identification considerations, and model extensions. Transp Res B 42(3):274–303
5. Bhat CR (2008b) Duration modelling. In: Hensher DA, Button KJ (eds) Handbooks in transport 1. Elsevier, Amsterdam
6. Bhat CR, Pulugurta V (1998) A comparison of two alternative behavioural choice mechanisms for household car ownership decisions. Transp Res B 32(1):61–75
7. Cameron AC, Trivedi PK (1998) Regression analysis of count data. Cambridge University Press, Cambridge
8. Chorus CG (2010) A new model of random regret minimization. Eur J Trans Infrastruct Res 10(2):181–196
9. Chorus CG, de Jong GC (2011) Modeling experienced accessibility: an approximation. J Transp Geogr 19:1155–1162
10. Cramer JS (1990) The logit model, an introduction for economists. Edward Arnold, London
11. Daly A, Bierlaire M (2006) A general and operational representation of GEV models. Transp Res B 40:285–305
12. Daly AJ, Zachary S (1978) Improved multiple choice models. In: Hensher DA, Dalvi MQ (eds) Determinants of travel choice. Saxon House, Farnborough
13. de Jong GC (1990) An indirect utility model of car ownership and private car use. Eur Econ Rev 34:971–985
14. de Jong GC (1996) A disaggregate model system of vehicle holding duration, type choice and use. Transp Res B 30(4):263–276
15. de Jong GC, Daly AJ, Pieters M, van der Hoorn AIJM (2007) The logsum as an evaluation measure: review of the literature and new results. Transp Res A 41:874–889
16. Geweke J, Keane M, Runkle D (1994) Alternative computational approaches to inference in the multinomial probit model. Rev Econ Stat 76:609–632
17. Greene WH, Hensher DA (2012) Revealing additional dimensions of preference heterogeneity in a latent class mixed multinomial logit model. Appl Econ 1–6
18. Hess S, Fowler M, Adler T, Bahreinian A (2009) A joint model for vehicle type and fuel type. Paper presented at the European transport conference, Noordwijkerhout, The Netherlands
19. Hess S, Ben-Akiva M, Gopinath D, Walker J (2011) Advantages of latent class over continuous mixture of logit models. Working paper, ITS, University of Leeds, Leeds
20. Holguín-Veras J, Xu N, de Jong GC, Maurer H (2011) An experimental economics investigation of shipper-carrier interactions on the choice of mode and shipment size in freight transport. Netw Spat Econ 11:509–532
21. Kahneman D, Tversky A (1979) Prospect theory: an analysis of decision under risk. Econometrica 47:263–291
22. Kahneman D, Tversky A (1992) Advances in prospect theory: cumulative representation of uncertainty. J Risk Uncertainty 5(4):297–323
23. Kristoffersson I (2011) Impacts of time-varying cordon pricing: validation and application of mesoscopic model for Stockholm. Transp Policy. doi:10.1016/j.tranpol.2011.06.006
24. McFadden D (1974) Conditional logit analysis of qualitative choice-behaviour. In: Zarembka P (ed) Frontiers in Econometrics. Academic Press, New York

25. McFadden D (1978) Modelling the choice of residential location. In: Karlqvist A, Lundqvist L, Snickars F, Weibull J (eds) Spatial interaction theory and residential location. North-Holland, Amsterdam
26. McFadden D (1981) Econometric models of probabilistic choice. In: Manski C, McFadden D (eds) Structural analysis of discrete data: with econometric applications. The MIT Press, Cambridge, Massachusetts
27. Miller EJ (2003) Microsimulation. In: Goulias KG (ed) Transportation systems planning: methods and applications. CRC Press, Boca Raton
28. Picard G, Gaudry P (1998) Exploration of a Box-Cox logit model of intercity freight mode choice. Transp Res E 34(1):1–12
29. Puckett SM, Hensher DA (2006) Modelling interdependent behaviour as a sequentially administrated stated choice experiment: analysis of freight distribution chains. Paper presented at the International Association of travel behaviour research conference, Kyoto
30. Rust J (1987) Optimal replacement of GMC bus engines: an empirical model of Harold Zurcher. Econometrica 55(5):999–1033
31. Significance (2009) Effect op autobezit van omzetting van de BPM in de Kilometerprijs. Report for the Environmental Assessment Agency, Significance, The Hague
32. Significance and Panteia (2012) Aanpassingen specificaties BasGoed V2.0. Report for DVS Rijkswaterstaat, Significance and Panteia, The Hague/Zoetermeer
33. Small K (1987) A discrete choice model for ordered alternatives. Econometrica 55(2):409–424
34. Tavasszy L, de Jong GC (2014) Modelling freight transport. Elsevier insights series. Elsevier, Amsterdam
35. Train K (1986) Qualitative choice analysis: theory, econometrics, and an application to automobile demand. MIT Press, Cambridge, Massachusetts
36. Train K (2003) Discrete choice methods with simulation. Cambridge University Press, Cambridge
37. van de Kaa EJ (2008) Extended prospect theory, findings on choice behaviour from economics and the behavioural sciences and their relevance for travel behaviour. Ph.D. thesis, Delft University of Technology, Delft
38. Vovsha P, Petersen E, Donnelly R (2003) Explicit modelling of joint travel by household members: statistical evidence and applied approach. Transp Res Rec 1831:1–10
39. Wen CH, Koppelman FS (2001) The generalised nested logit model. Transp Res B 35(7):627–641

# Collaborative Assessment Processes

# 9

Ilaria Tosoni and Felix Günther

**Abstract**

Large-scale spatial strategies clearly belong to the category of complex planning problems [15–20, 14, 18]. They represent a challenge in terms of technical competences and administrative effectiveness, but offer also a very interesting chance for experimenting with collaborative planning instruments. Collaborative environments [2] are necessary to support the creative emergence of proposals and ideas best corresponding to the construction of a consensual and integrated vision of the future among different stakeholders, in which the settlement development, infrastructure and management policies become the structuring elements. The central question then becomes how to organize these processes of discussion and joint planning and which tools can be put in place to facilitate the establishment of patterns of collaborative interaction in these large-scale spatial contexts. The chapter focuses on the search for a working method to tackle these issues and reports on the experiments conducted within the framework of the CODE24 Interreg projects. The approach developed sees in the practices of joint assessment the central moment of spatial strategies' design. In particular, the Collaborative Assessment process is presented as key instrument focusing on a dialogical mapping of the issues at stake and on accompanying the key spatial actors into the definition of a desired development horizon through a process of empowerment of their design skills.

I. Tosoni (✉) · F. Günther
ETH Zürich Institute for Spatial and Landscape Development, Chair for Spatial Development, ETH Zürich, Wolfgang-Pauli-Street 15 (HIL H 37.4), 8093, Zürich, Switzerland
e-mail: tosonii@ethz.ch

F. Günther
e-mail: fguenther@ethz.ch

I. M. Lami (ed.), *Analytical Decision-Making Methods for Evaluating Sustainable Transport in European Corridors*, Sxi 11, DOI: 10.1007/978-3-319-04786-7_9,

## 9.1 Introduction

The magnitude of problems faced in the practice of spatial development processes, increasingly embraces a large-scale territorial dimension. Examples are policies of development and management of large metropolitan areas or urban regions, as well as, for example in the Swiss case, the integrated planning of valley areas. In particular, to this category belong large infrastructure projects, which physical and functional impacts concern territories at the interregional and, in the European case, often also international level. These are issues of strategic importance, which need to be dealt with in an appropriate way, a great planning effort and accompanying processes mobilizing skills, interests and resources spread in the territories themselves.

Many different experiences and processes can be analysed aimed at this type of project development showing very different settings and outcomes [1, 4, 13]; however, a proper formulation of methods and instruments, able to deal with these kinds of situations, it is not yet available as part of planning and spatial development practice.

The research group of the ETH Zurich, which refers to the Chair of Spatial Development, has been experimenting in recent years with these issues, trying to specify a working method, allowing to inserting the design of spatial strategies within collaborative decision-making processes. The testing involved different geographical areas: for extension, features and issues addressed. An important area of work and experimentation has been the macro-region described by the so-called European railway corridor Rotterdam—Genoa. In this area it was possible to design a process that would lead to the definition of a regional strategy developed by the regions and major stakeholders along the corridor. This process has taken on in 2010, the form of a European project (CODE24), which, over a period of 4 years, has been an important laboratory for testing of a method of working and new tools accompanying the different stages of elaboration of territorial strategies. The processes of collaborative assessment, according to the method developed, focus on a dialogical mapping of issues, in their essential parts, providing, as far as possible, their spatial representation and accompanying the actors in the definition of a desired development horizon through a process of empowerment of their design skills. The approach developed sees in the practices of assessment, the central moment of these spatial strategies design, which mainly affects two phases of the work:

- The phase of evaluation and discussion of problems and conflicting interests.
- The stage of comparison of design alternatives and their expected impacts.

In these crucial steps, the difficulty lays in simultaneously assessing factors and elements of different nature and different scope and reaching a satisfactory synthesis, in terms of technical feasibility, operational functionality and consensus. The assessment, in fact, does not only concern technical and design issues, but always implies a judgment on priorities, which derives from evaluations depending on the training sector, role, or beliefs, hierarchical position, etc. of the actors. In collaborative processes is therefore important that these references emerge in order to highlighting the non-neutral nature of problems and solutions definition.

## 9.2 Complex Problems and Ad hoc Instruments

The search for a working method originates with the inclusion of large-scale spatial strategies in the category of the problems of complex nature [15–20, 14, 18].

In the school of planning known as Action planning [15] complex problems are defined [15] as: *comprehensive and spatially relevant tasks, in which it is not clear through which projects they can be solved. They also cannot be solved through a sectorial approach, but ask for over-reaching professional and organizational cooperation, often also between different levels of government* [15, p. 36].

These are situations in which different levels of investigation and critical issues add up and their interpretation clearly depends on the perspective taken by the various parties involved, which may involve simultaneously: perceptions about the nature of problems; support to specific technical solutions; the protection or development partisan interests; the achievement of political or professional goals etc. For example, in the case of the design of an infrastructure project of strategic importance some stakeholders will focus more on the local scale impacts (noise, increased traffic, outages, etc.), others on technical and design issues, others on the symbolic or functional meaning of the work. Bringing these different views to converge on a shared project proposal is often a difficult challenge, which must be matched by adequate organizational effort. The Action Planning postulates, in fact, that complex problem require for their solution the application of ad-hoc instruments [10, 15] designed to fit the specific context and situation addressed. The uncertain and diversified nature of such situations also implies that to achieve a robust design solution; several attempts and thus different solutions and instruments are often necessary [10]. Through this circular path, it is the very nature of the problems to be better interpreted and understood, and not only the solutions. These issues become more relevant in the present context characterized by an increasing fragmentation of interests and lack of players able, out of acknowledged authority, to guide the decision-making. The definition of the problems is, therefore, primarily a process of social construction of meaning [19, 20], which cannot do without the interaction between the various stakeholders affected by territorial transformations. These processes therefore need to be accompanied and designed in order to create the collaborative environment [2] useful to the emergence of proposals and ideas that best correspond to the construction of a shared vision of the future in which the settlement development projects, infrastructure and management policies become the structuring elements. The central question then becomes how to organize these processes of discussion and joint planning and what tools can be put in place to facilitate the establishment of patterns of collaborative interaction.

As part of the Action Planning, through a series of experiments located in different territories, mostly at the local scale, the Test Planning procedure [15] has been developed over the last 20 years, as main instrument for facilitating the resolution of spatial problems of a complex nature. It is a procedure for collaborative planning based mainly on the involvement of expert knowledge. They are involved in the process in different ways, through the assignment of a specific role: moderator, the technical secretariat of the procedure, designers, representatives of local institutions, external experts, etc. Core element of the procedure is to provide for a tight scheduling of times, which usually is articulated over 6/8 months, in order to avoid the dilation of the discussion and consequently the risk of non-productivity of the process and decision-making.

The Test-planning has proven to be an effective tool for the understanding of complex local situations [20, 5]. The main limit of these experiences lies in the episodic character that characterises them: they are moments, by their nature, separated from the routine of the parties involved and which is usually limited to the design phase, without offering a phase accompanying the course of implementation and transposition of guidelines and recommendations in the ordinary tools of planning. Moreover, in many cases the contribution of design is purely instrumental and rarely it is possible a real stakeholder involvement in the development of the elements of the intervention strategies. These aspects are not suitable to achieving a central goal of the construction of spatial strategies: that of creating a long-term partnership that can go beyond the limits of single planning episodes, to substantiating a mode of constant interaction, to be activated in several occasions and with respect to different objectives.

## 9.3 The Process as a Peer-Learning Path

The different experiments carried out, particularly in the context of the project CODE 24, made it possible to initiate a reflection on the instruments best able to perform the essential task of building a partnership in support of spatial strategies. The instruments have been progressively refined, starting from the Test Planning experience, and incrementally taking cues resulting from the different experiments [9]. In particular, it has been consolidating over time the perspective that sees the collaborative process of constructing spatial strategies as peer learning processes [3, 8, 21]. They essentially are moments of sharing and re-elaboration of knowledge and skills possessed by the different actors involved in the designing process, which, however, in order to be useful, need to be fertilized and activated by the interaction and the transposition into concrete intervention proposals.

In these processes, the problem definition is done in a dialogical way from a first definition, which is at first conveyed by one of the parties, and which is progressively deepened starting from different conceptual contributions. The starting point is therefore a stimulus, an initiative of the promoter that initiates discussion and brings together around the theme of the attention of other bearers of knowledge. In the case of large-scale regional strategies two aspects are crucial:

- The form of interaction.
- Capacity building of design skills.

The form of interaction, and therefore the kind of organization chosen to support the process, strongly influences the ability to build a productive working group [15]. To this purpose is crucial to provide different moments of interaction characterized by different organizational settings. As expected, processes affecting significant areas, such as that covered by the Rotterdam Genoa corridor, require the participation of a large number of actors. In this case, for example, considering only institutional actors, it is easily possible to reach more than 200 subjects (states, regions, municipalities, state enterprises, etc.), which can grow exponentially when including the various stakeholders (railway operators, logistics companies, associations of citizens). It is clear that these numbers are difficult to manage productively through a reduced number of meetings or in plenary. In addition, within these groups, there are a variety of composite roles (for rank, specific skills, attitude to discussion). It is therefore necessary that the different work settings also promote free moments of interaction where all actors can contribute regardless of their competences and specific functions. In addition to this there are the obvious cultural differences involved acting in interregional and international contexts:

- Linguistic differences.
- Approach to interaction.
- Different influences disciplinary.
- Differences resulting from the different contingencies.

Also in this case it will be necessary to bring out the cultural or "local" aspects implied by the different formulation of problems in order to focus the design interaction. The settings identified for the management of these decision-making contexts are manifold:

- Forums and plenary presentations that include ex cathedra presentations and discussion.
- Debate in thematic sessions.
- Work in small groups with the help of facilitators.
- Work in sessions of structured evaluation, starting from previously processed proposals.

Within the project CODE24, as will be better explained later on, it was chosen to modulate the different options, so as to ensure both the widest possible communication and discussion, and the possibility of in depth thematic analysis; always with the aim of reaching the drafting of concrete proposals for action.

## 9.4 The Project Element and Design Capacity Building

Another fundamental aspect emerged from the application of different instruments concerns the centrality of the design activity as part of the collaborative strategy definition. In many cases, in fact, the discussion in these contexts stops at policy level or general criteria the intervention, which frequently take the form of exhortative slogans. Progress is rather produced when, from the discussion on principles, it is possible to shift towards the debate on project components themselves. This step allows shifting the focus of attention to very specific and concrete contents. The operation, of centering the focus of the discussion back to space, sees the project as his medium of choice. It is, however, an operation that requires specific skills. Although many of those involved in this type of processes have the technical expertise to deal with these questions, the various processes accompanied so far tell us that it is very unlikely that participants will spontaneously undertake such effort. This depends on many factors: poor habit to try their hand with the project, willingness to respect the dynamics of the group, but also fear of judgment and of the immediate personal exposure resulting from the presentation of a conceptual contribution. This aspect refers, on the one hand, to the need for a setting fostering the possibility for multiple parties to contribute in a free and protected way, and, on the other hand, shows the strength of the learning design time, which let latent contradictions and unresolved issues emerge. Moreover it should be pointed out that in the context of processes of large-scale strategy design, even in technical terms, there is a lack of experience and tools that are available to the different actors to translate strategic contents into planning representations. In these cases the role of facilitators is therefore about developing a new design language, adapted to the different scales of the topics covered, and a mode of communication and analysis, allowing the simultaneous participation of subjects carrying distinctive knowledge and interests. Design is therefore a powerful tool at key stages of the assessment process:

- At the beginning it enables the discussion on the contents.
- It becomes active ground of negotiation and creation of new knowledge.
- It is the main tool of communication.

## 9.5 Instruments Supporting Collaborative Planning: The Collaborative Assessment

Trials on these matters have been possible in recent years (2009–2013), as part of several projects, which have seen the research group of the Chair Spatial Development of ETH Zurich engaged in accompanying processes of definition of large scale territorial strategies [11, 17]. In particular, the project CODE 24 represented a true laboratory for testing and checking the theoretical reflections carried out, within real processes and through different settings of interaction between the actors. The starting point of the project was the acknowledgment by the regions involved, of the strategic nature of the definition by the European

Union of a rail infrastructure corridor linking the ports of Northern Europe (Rotterdam, Antwerp, Hamburg) directly with the terminal of the northern Mediterranean (in this case represented by Genoa). From the point of view of the regions that choice presents itself as an opportunity for development, but at the same time as potentially problematic issue due to the strong mission of freight axis, which national and European decision-makers foresaw for this the corridor. It is therefore clear that, in terms of regional spatial development policies, this goal produces significant planning effects: both in terms of ability to effectively seize the opportunities offered by the infrastructure development (through improving accessibility; services upgrade or creation of new services; territorial marketing), but it also demands intervention aimed at granting the overall sustainability of the expansion of such infrastructure in densely populated areas and bearers also of other instances potentially in conflict with the foreseen use. On the one hand, the issue of noise and other emissions control emerges more and more virulently, as a compensation factor and achievement of acceptable levels of quality of life (especially in some regions), and on the other discussions arise about the compatibility of this strategic plan with the need to strengthen and improve the quality of both the suburban rail services in large metropolitan areas, and the connection between the different metropolitan areas themselves, through efficient long-distance services. It is clear that, in this case, the need for integration and coordination between economic, infrastructure operational and spatial planning objectives arises in order to allow a coherent strategic plan for the future of the territories involved. In addition to these challenges, it must be added the international dimension and therefore the range of subjects and issues involved.

Within the project CODE 24, in fact, act in various capacities: Supranational bodies; Six states with their national ministerial representatives; Eight regions NUT2 and 18 NUT3 regional authorities; Municipalities and metropolitan areas; Citizens' associations; Railway operators; Port authorities; Business Companies; Logistics operators. From this point of view it is possible to understand the interest for the possibility of research and experimentation offered by the project CODE24 and the challenge it represents. The method of intervention developed, which finds in the collaborative assessment his main instrument, is therefore based on the aspects previously mentioned:

- Encourage interaction among participants.
- Foster a long-lasting collaboration.
- Encourage the setting of shared and discussed criteria for the evaluation of issues and policy alternatives.
- To promote concrete intervention actions through the use of project design.

The path devised aims at encouraging, in particular, two key aspects of the participation of the actors:

- The introduction and enhancement of their knowledge about the territory in the process of issues' emergence.

- The facilitation of their contribution, not only to the evaluation of design alternatives, but also to develop own proposals.

Following several trials, the path has been consolidated into four phases and modules:

- Mapping and reconnaissance of the main issues at the local level and their reconstruction and representation.
- Expansion of the scope and comparison with other regions; construction of an overview of emerging critical issues arising from the failure of policy coordination.
- Project development prior among experts and later open to all stakeholders.
- Joint evaluation of alternatives, through collaborative design and synthesis of the relevant issues.

Through this scheme, repeated in several cycles and regions in an incremental manner, it is possible to arrive to a reasoned and critical mapping, because investee, of the various components that inform the spatial strategy and to a shared definition of the guiding principles of the strategy itself.

## 9.6 Applications at Different Scales

The proposed method has been applied both with the goal of building a shared strategy for the entire corridor and working on inter-regional cooperation at a smaller scale, in areas presenting particular problems. In particular, the working method was tested, in a first phase, in the metropolitan region of Frankfurt and Mannheim and later extended to the entire corridor. The two paths are, however, complementary and part of a single process.

As already explained, the process of collaborative assessment has been defined through the experiments implementing an adaptive strategy: instruments and modules have been gradually adjusted and re-set through the interaction with the different stake-holders. In its final outlook the collaborative assessment process consists of four modules of intervention with different purposes, which are accompanied by suitable instruments aimed at establishing different modes of interaction and involvement [12].

In the CODE24 experiment, the first module of mapping has affected all regions along the corridor through the organization of local focus groups (max 7/10 participants). It has been a series of one-day meetings, which involved in priority local partners of the project, with the support of local experts (selected by the partners). The goal has been to build together a first representation of the problems and opportunities based on the participants' knowledge of their territory. The second level of exploration then concerned existing projects: proposals having directly to do with the design of the corridor (e.g. work on the rail network or concerning the) or which, for whatever reason, relate to it (new housing projects, logistics centres, functions or services that need great accessibility, etc.). The mode of interaction has been that of the discussion starting from questions prepared by the organizers and feeding of the information by the participants. The summary of such meetings, by the researchers, includes a narrative illustration of the topics, but also their first visualization in the form of overview maps.

These latter materials has then become the starting point for the discussion in larger groups (30–50 people), always on a regional scale, but also introducing the theme of the corridor as a whole and in particular in relation with neighbouring regions. To this purpose therefore actual workshops have been organized and structured in three phases:

- A frontal introduction of key themes made by the experts previously involved and by the research team, which shows the contents emerged in the focus groups.
- Thematic sessions in small groups led by a facilitator.
- Discussion and concluding summary.

This activity identifies the second module of intervention and has been applied too in each of the regions involved in the project.

The work of reassembling all elements together in a single map was finally carried out by the researchers, based on the key themes emerged in the various discussions. This first product, named first assessment of the situation, was submitted to the project partners and discussed during several project meetings. The module introduces a first planning character, because the processing aims at selecting geographical areas, which require specific design by differentiating between:

- Local areas where intervention is required to re-design the infrastructure.
- Supra-local areas in which the project requirement concerns the reconstruction of a shared strategic horizon due to the overlap of multiple design themes. In these cases, the infrastructure can have a strategic role only if effectively inserted in an overall strategic plan.

The initial input (by the researcher) has the aim of enabling a design-oriented discussion to take place, by providing the participants with a common visual language and simple information packages to tackle the strategic planning task.

The discussion sprung from the joint examination of this first evaluation, in fact, allowed to identifying four macro-areas of intervention:

- The Mittelrhein: where an in-depth solution-design process should propose cost-benefit (efficient) solutions to the environmental problems.
- The Frankfurt-Mannheim region: where an efficient improvement of all networks could be achieved thorough flows separation.
- The Oberrhein region: where a better coordination between the Rhein network and the French network could be highly beneficial together with a smart solution for the gateway of Basel.
- The Ligurian range: where the empowerment of the port of Genoa needs to be achieved through a gradual but intensive investment plan.

**Table 9.1** Collaborative assessment: the four modules

| Module | | Task | Mode of interaction | Instrument | Design object and technique | Scale | Number of experiments | Number of participants | Language |
|---|---|---|---|---|---|---|---|---|---|
| **Exploration** | Problem setting | Mapping | Structured discussion | Focus groups | Thematic overviews<br>Thematic Maps and Charts | Regional | 12 | Ca. 70<br>Project Partners + External Experts | German<br>Italian<br>Dutch<br>English |
| | Emergence of opportunities | Mapping | Structured discussion, Brain storming | Focus groups | Potential areas of intervention<br>Geographical and symbolic maps | Regional | 12 | Ca. 70<br>Project Partner + External Experts + Local stakeholders | German<br>Italian<br>Dutch<br>English |
| **Identification of interdependencies** | | Assessing key information | Input and reaction<br>Open discussion | Workshops | Areas and project of strategic interest<br>Spatial conflicts<br>Geographical and symbolic maps | Interregional | 15 | Ca. 400<br>project partners + external Experts + local and external stakeholders | German<br>Italian<br>English |
| **Assessment of the situation** | | Strategic comparison of the information: definition of work and intervention areas | Input and reaction in several feedback cycles | Presentations | Experiment of synthetic visualizations: geographical and symbolic maps; specific project proposals | Macro-regional (whole corridor) | 3 | Ca. 40<br>Project Partners | English |
| **Collaborative Assessment** | | Joint evaluation of complex planning situations and the elaboration of intervention proposals | Structured discussion, Simulation, Role playing, Open discussion | Workshops<br>ANP strategic assessment<br>Strategy cards | Experiment of synthetic and dynamic visualizations: geographical and symbolic maps; real-time interactive visualizations (InViTo) | Macro-regional (interregional and whole corridor) | 4 trial tests<br>2 workshops | Ca. 40<br>Project Parters + External Experts and researchers + Stakeholders | English<br>German |

([*Source* Tosoni, Guenther, ETH Zurich])

A priority setting processing on both regional and corridor's scale was then the object of the fourth activity: the collaborative assessment workshop. The idea behind the module is that the design of these spaces cannot be left to the national or regional authorities alone, but it should be attended in a collaborative process involving all planning levels and authorities and important stakeholders.

In order to allow this process the collaborative assessment procedure has been designed as an instrument accompanying the evaluation of complex planning situations and the elaboration of intervention proposals in a collaborative way. The procedure has been first tested at the inter-regional scale and later further developed to be adapted to the corridor dimension (Table 9.1).

### 9.6.1 A First Experiment: Frankfurt-Mannheim Region

According to the information collected and discussed during several focus groups and workshops with the stakeholders (five in total for the two regions), it emerged the strong inter-dependency between the two metropolitan areas, which functionally already work as singular macro-region. The collaborative assessment workshop has then been designed in order to promote the dialogue between representatives of both regions and support them in designing a common strategy. The main issues affecting this conurbation have been identified by the stakeholders through the mapping sessions and can be reported as follows:

- Rail infrastructure and operation: The Mannheim-Frankfurt route is one of the busiest routes in the North-South axis, with a total of nearly 700 trains on the three parallel double tracks. The three routes have a specific individual function: long-distance traffic dominates on the Riedbahn, freight transport and regional services on the Main-Neckar-path, on the Bergstrasse and on the Rhine route. Nevertheless mixed traffic is offered on all routes and this of course severely limits the network capacity.
- Key project under discussion is the new high-speed traffic route Frankfurt-Mannheim. This new line is supposed to loose and eliminate all the bottlenecks that are now recognized on the network. The proposal for this fast connection aims at disengaging capacities on the normal routes to the south in order to redirect also there the intensification of settlement location.
- Spatial development: The area of Mannheim is subject to 'positive stagnation': a slow reduction of the demographic development. The settlement development is characterized by immense reserves, on the one hand, fed by the industrial conversion and de-localization, on the other hand, by the withdrawal of the U.S. military. This combined with the shrinking of the population leads to great opportunities and weak demand. The ports need to promote cooperation in order to better exploit their capacity. A strong competition among the different uses puts pressure on port and logistics areas and pulls the development towards uses with greater added value. The same situation regarding the port can be found also in Frankfurt, nevertheless with a much higher pressure on industrial

conversion due to the dramatic demand for settlement areas in the region. Such increasing pressure is producing a strong suburbanization of the region (specially in direction of Mainz-Wiesbaden and at the access points to the infrastructure network). The development takes place at a high rate apparently without the needed coordination.

A crucial phase in the process has proven to be that of switching from the analysis stage to the design phase. The participants in all settings found having to work in a conceptional way extremely compelling and demanding. This does not only relate to the usual working habits of the stakeholders in questions, but also to the level of *social exposure* connected to the addressing of ideas and project proposals, especially about such complex situations and in the framework of experts' communities. In order to support this transition three alternative strategies have been designed in order to allow maintaining the already experimented assessment modus, but introducing a stepwise design component. The participants were asked not only to compare the strategies, but to identify the key components and try to use them to compose a best fitting option in alternative to the previous three. The strategies were designed together with a small group of local partners in two design workshops. They have been characterized as extreme development perspectives embodying the view of possible alternative target groups (representatives of the knowledge economy; regional and local planners and interest groups; forwarders and logistics operators):

- Frankheim: a new cross border metropolis. The strategy focuses on the development of a new high speed line connecting the 2 main cities of Frankfurt and Mannheim in about 20/25 min.
- NET-Regio: a regional agglomeration. The strategy focuses on the improvement of the existing transport network. It includes a set of small investments on the different lines, which can produce a significant improvement in the capacity along the lines by separating long distance form freight and regional flows.
- EURO Hub: the European logistic platform. The strategy focuses on the implementation of an integrated logistic concept for the region. Main intervention will be the realization of a railway line, dedicated to freight transport: the line will allow the realization of additional tri-modal logistic hubs providing a combination of services and areas (Fig. 9.1).

The assessment took place in two separate workshops involving project partners and external experts. The analysis of the strategies was conducted using both quantitative (ANP) and qualitative methods (Strategic assessment). Key feature of both has been the deconstruction of the strategies in their basic elements and attributes. Though the comparison of couples of elements in combination to their visualization [9], the discussion has been productively guided to letting core discussion topics emerge. Such discussion aimed at identifying not only the preferred strategy, but also the key components of a desired one. From the joint selection of these main issues an in-depth exploration of further implication of the three

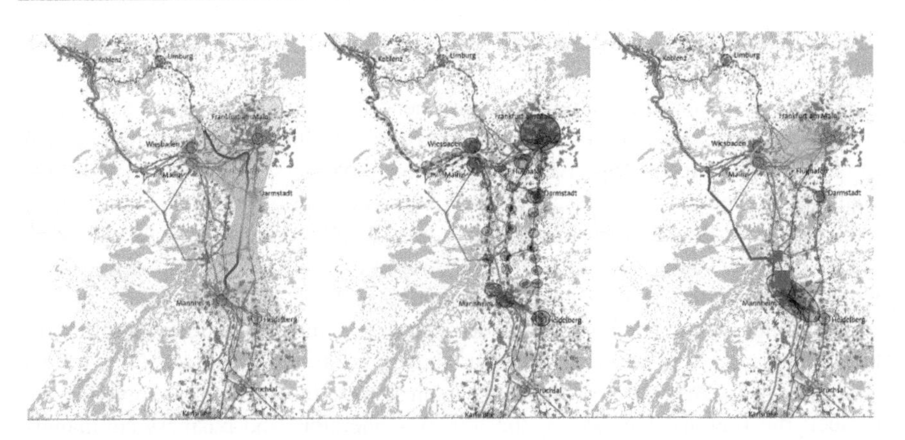

**Fig. 9.1** Frankfurt/Mannheim 3 spatial strategies: Frankheim, SuperSBahn, EuroHub [6] (*Source*: Tosoni, Guenther, Eth Zurich)

strategies was conducted (Strategic Assessment: Guenther, Tosoni 2011 [6]) in smaller groups. Aim of this activity has been to focus on the spatial impacts of the chosen development path and to envisage possible betterment of the proposed solutions. In the final session the participants were asked to share the outcomes of the groups' discussion. In this crucial phase, facilitators supported the process by underlying the elements emerging as consolidated solutions by the participants, and the fields where the debate is still open and further intervention is needed.

The whole process lasted about 6 months and involved a core group of about 12 people (researchers and local partners). To the two official workshops participated 20 stakeholders, representing local institution, regional authorities, business companies and also national authorities. The experiment concerned both the contents and the instruments used. It has been very important to focus on way to support the participants' familiarization with the definition of the strategy's components and with the different modes of visual representation, in order to provide them with tools to steer interaction. Therefore the approach has been based on the gradual delivery of information collected and re-elaborated by the researchers through different media: face-to-face meetings, maps, reports, on-line questionnaires and videos. All these instruments have been constantly updated and upgraded in order to be able to swarm the inputs of the participants and support a transformative process of joint knowledge production. The main purpose of the different tools was in fact to transform simple packages of information into shared knowledge through the direct interaction between the participants. The group of participants developed with time an own view on the issues and was able to use the tools provided to elaborate a shared conceptual framework. This learning and empowerment process enabled the gradual establishment of a permanent working-group able to activate other regional and national authorities involved in the formal decision-making, but also engaged in adopting the common framework as a reference in their routine planning activities.

### 9.6.2 Corridor's Assessment

Following the Frankfurt-Mannheim experiment, the next task has been to focus on structuring the assessment for the whole corridor. This task was the central activity of the CODE24 project and involved both project partners and invited experts. The challenging element of these tasks concerns the amount of information and inputs to be considered in order to approach the topic of a common strategy for the whole macro-region. The preparation of this discussion took about 6 months during which several internal trials have been conducted.

As for the local workshops competitive strategies have been developed in order to support a content-based discussion: the three strategies reflected, in a very simplified way, the views of the three main stakeholders groups involved in the project: the regional authorities, the logistics operators and business companies, the environmentalists and the citizen's associations.

The assessment has then been carried out using different approaches:

- Through visualization: Comparing alternative strategies.
- Through discussion: Defining the key elements of a possible strategy.
- Through gaming: assembling a strategic proposal starting from a set of given project-packages.

In this case too, the result was a set of preferences concerning the key elements of a common strategy, but most importantly the definition of a path for designing a permanent collaboration platform. Aim of this platform should be to enable long-term collaboration between the regions sponsoring projects and common activities to supporting sustainable development along the corridor. Among the main outcome of the assessment, in fact, there has been the formulation of a work plan for the subsequent (and operational) stages of maintenance of the partnership. Already in the formulation of the Interreg project it had been detected, in the formation of an EGTC (European Grouping of Territorial Cooperation), the possible platform to keep the future cooperation between the partners active. European legislation in fact offers the possibility of forming a legal entity which may be delegated by the partners underwriters functions of common interest. However, it was through the collaborative assessment that the possible tasks of the EGTC were better defined given concrete form.

## 9.7 Towards a Method for the Management of Collaborative Processes at the Large Scale

The experiments carried out, mainly through the project CODE24, allowed to consolidate a working method and an instrument, the collaborative assessment, which have proved productive in assisting the definition of large-scale territorial strategies. The assessment, designed in 4 modules, allows initiating and accompanying collaborative discussion between the different stakeholders and managing potential conflicts through negotiation. In this task it is assisted by the use of the design, which is offered as a tool for exploration and verification of shared design solutions. Some elements are crucial in this process:

- Flexibility of the instruments and adaptation to different contexts: through replicable, but modelled on different situations work modules.
- Focus on communication and interaction among peers: providing different work settings.
- Engagement of the stakeholders by modulating commitments: light, but constant in the long run, through forms of remote collaboration and intense and engaging on the occasion of the workshops.
- Incremental path toward designing: through different forms of visualization and attempt with the design practice.

In addition, the experiment carried out during the Collaborative Assessment Workshops, through the inclusion of quantitative and structured evaluation methods for ranking design alternatives and components, opens a new and interesting area of research. In fact, it concerns the development of tools to support the assessment, both with regard to the construction of shared criteria for evaluation, and with respect to the issue of the language of spatial representation.

Finally time represents to all effects a strategic variable of fundamental importance: as in the case of the Test Planning [11, 15], the transition phase between the assessment workshop event and the ordinary planning routine is a delicate and crucial step for the survival of the spatial strategy. In the case of collaborative assessment, the direct participation of the stake-holders to the definition of the strategy's components (and therefore to the learning process and the construction of new knowledge that accompanies it) seems to guarantee a solid commitment and involvement of the actors in the long run. However reinforcement and follow-up interventions are recommended and advocated anyway. This also means that although it is possible to achieve results with a one-day workshop, summing up the incubation, preparation and follow-up stages of the assessment procedure, the timing of development of the process are still quite long and therefore they as well require adequate financial resources to fund the different phases. Even in this respect the constitution of a subject (such as the EGTC), responsible for the organization and maintenance of this type of activity at the macro-area level, could be the appropriate instrument to take over the university in the transition from the experimentation to consolidation phase of the practice.

## References

1. Albrechts L (2006) Problems and pitfalls in large-scale strategic planning. International Congress "La Citta di Citta", 20 Feb 2006, Milano
2. Axelrod R (1984) The Evolution of Cooperation (Revised ed.). Perseus Books Group
3. Budge K, Beale C, Lynas E (2013) A chaotic intervention: creativity and peer learning in design education. Int J Art Des Educ 32 (2):146–156

4. Deppisch S (2012) Governance processes in Euregios: evidence from six cases across the Austrian—German border. Plann pract Res 273:37–41
5. Elgendy H (2003) Development and implementation of planning information systems in collaborative spatial planning processes. University of Karlsruhe, Dissertation, Karlsruhe
6. Günther F, Tosoni I (2011) Il progetto Interreg code 24: strumenti e azioni per la cooperazione interregionale lungo il corridoio Rotterdam—Genova. Paper presented at the INU—IX Biennale delle città e degli urbanisti europei, Genova, Italy, 14–17 Sept 2011
7. Healey P (2002) Place, identity and local politics: analysing initiatives in deliberative governance. In: Hajer M, Wagenar H (eds) Deliberative politic analysis: understanding governance in the network society. Cambridge University Press, Cambridge
8. Klein G, Klug N, Todes A (2012) Spatial planning, infrastructure and implementation: implications for planning school curricula. Town Reg Plan 60:19–30
9. Lami IM, Masala E, Pensa S (2011) Analytic network process (ANP) and visualization of spatial data: the use of dynamic maps in territorial transformation processes. Int j Analytic Hierarchy Process 3:92–106
10. Maurer J (2005) Planerische strategien und taktiken, ARL
11. Nollert M (2013) Raumplanerisches entwerfen, entwerfen als schlüsselelement on klärungsprozessen der aktionsorientierten planung—am beispiel des regionalen massstabs, Dissertation, ETH Zurich
12. Otsuka N, Günther F, Tosoni I, Braun C (2014) Developing trans-European railway corridors: lessons from the CODE24 project, submitted to Transport Reviews
13. Pickering T, Minnery J (2012) Scale and public participation: issues in metropolitan regional planning. Plan Pract Res 27:249–262
14. Rittel H, Webber M (1973) Dilemmas in a general theory of planning. Policy sciences, Vol 4, pp 155–169
15. Scholl B (1995) Aktionsplanung. Zur behandlung komplexer schwerpunktaufgaben in der raumplanung. vdf-Hochschulverlan der ETH, Zürich
16. Scholl B (2005) Strategische planung. ARL: 1121–1129
17. Scholl B (2012) Gedanken zur nord-sued-Transversale, TEC21. 17:31–33
18. Schönwandt W, Voigt A (2005) Planungsansaetze. Handwörterbuch der Raumordnung. ARL, Hamburg
19. Schönwandt W (2011) Grundriss der Raumordnung und Raumentwicklung. ARL
20. Schönwandt W (2013) Solving complex problems. Jovis
21. Smith B, MacGregor JT (1992) What is collaborative learning? publisher: national center on postsecondary teaching. Learn, Assess Pennsylvania State Univ 117(5):10–30

# Visualisation: An Approach to Knowledge Building

# 10

Elena Masala and Stefano Pensa

**Abstract**

The chapter proposes the use of the Interactive Visualisation Tool (InViTo) as a method for sharing information by using spatial data visualisation, also known as geo-visualisation, applied to support spatial decision making and planning. InViTo is based on the idea that interacting with data can improve the users knowledge process, while visualisation should contribute to increase intuitive perception. Therefore, through the interactive and visual exploration of geo-referenced data, participants to spatial processes are supported to evaluate strategies and objectives for several alternative development options. The visual system works both on two-dimensional and three-dimensional views, so as to better meet users' skills in interpreting images. InViTo has been used in different applications, with diverse purposes and spatial scopes, showing its effectiveness in creating a common language among the involved actors and enabling discussions on spatial development.

## 10.1 Introduction

Visualisation is a discipline which studies how to translate data and information into knowledge by means of visual communication [25–27]. In the scientific field, it studies the visual language in order to use human perception as a means for knowledge building.

---

E. Masala (✉) · S. Pensa
SiTI—Higher Institute on Territorial Systems for Innovation,
via Pier Carlo Boggio 61, 10138, Turin, Italy
e-mail: elena.masala@polito.it

S. Pensa
e-mail: stefano.pensa@polito.it

I. M. Lami (ed.), *Analytical Decision-Making Methods for Evaluating Sustainable Transport in European Corridors*, Sxi 11, DOI: 10.1007/978-3-319-04786-7_10,

Visualisation has grown into a computer-based field which offers a method for seeing the unseen [15], overcoming representation of pure data and deepening the understanding of the relationships among data. In that, visualisation is not just a visual translation of numbers and texts, but an organised framework of connections among different elements which allows information to be located and understood.

Nevertheless, when dealing with complex systems and huge amounts of data, visualisation can reach high levels of elaboration, showing multifaceted aspects and intricate relations among elements. In cases which comprehend all the geographical issues considered by spatial and transport planning, visualisation can propose visual representations with different levels of possible understandings and interpretations. On a lower level, visualisation can be used for generating useful images to share information among a large audience, while at a higher level, it can provide very complex and detailed information to specific experts. In this case, visualisation becomes a way to acquire knowledge, which can be enhanced by the exploration of data and the discovery of all the hidden connections which tie the various elements to each other. This means that visualisation should allow the contents to be investigated and understood whilst providing a dynamic framework for obtaining the information.

Literature on geovisualisation and geovisual analytics covering this issue suggests the combined use of visualisation and interactive tools [1, 2, 9, 10]. In fact, unlike static visual representations, interaction can introduce a dialogue between data and user, who can formulate specific queries and receive the relative feedback. This exchange of questions and answers provides important opportunities for the users to investigate the data and elements embedded within the visualisation, so that they can obtain richer information and build their own knowledge reaching high awareness on the represented issues.

Therefore, a high level of interaction represents the way for data exploration and knowledge building [9], generates a method which enables the cognitive process of large amounts of data and their complex relationships. Of course, this process is more successful if the users are experts in the considered field.

For these reasons, this chapter will discuss the use of interactive visualisation as a method for supporting the decision-making process within the planning process. It will present the Interactive Visualisation Tool (InViTo) as an instrument for enhancing the debate during collaborative processes and for contributing to the understanding of the debated subject. Finally, the theoretical concepts that lay beyond the technical proposals will be illustrated, showing their developments along the CODE24 project.

## 10.2 Interactive Visualisation: A Way for Involving Decision-Makers

Interactive visualisation is commonly considered as the right combination for knowledge building and implementing the users' awareness of the represented data [2, 6]. In particular, practitioners in urban planning ask for interactive tools to support spatial decision-making processes [24]. In this sense, technologies are developing fast and a wide variety of new instruments are nowadays available.

Nevertheless, the importance of using interactive tools within spatial decision-making processes is mostly underestimated by technicians working in the field of spatial issues. Generally, the computer-based research in spatial studies is not met by the daily needs of practitioners [4, 7, 8, 19, 20, 24, 28]. Many studies focus on the improvement of automatic processing applied to specific tasks such as analysing spatial data or simulating future scenarios, which are all activities that commonly require a long time for calculating results. Although these calculations provide the opportunity to receive detailed answers on specific questions, the models beyond these calculations are too complex to be easily understood by non-technical users. On the contrary, the requests of practitioners converge on simpler models, which could increase the interest and improve the cognitive process of the actors involved in the planning or decision-making practice.

For these reasons, the research on Planning Support Systems (PSS) and spatial Decision Support Systems (sDSS) have considered the interactive visualisation as a possible methodology for enabling new approaches within collaborative planning processes. In fact, while visualisation enhances the intuitive skills of participants, the interaction can speed up the acquisition of information, thus largely contributing to the individual's knowledge building. The fruit of this research is the Interactive Visualisation Tool (InViTo), a visual methodology for organising, investigating and exploring data and their imperceptible connections [17, 18, 21].

### 10.2.1 InViTo: Concepts and Framework

InViTo is based on two assumptions. Firstly, it supposes that the successfulness of a decision-making process strongly depends on the possibilities of communication between the people involved. Secondly, it assumes that a decision-making process can be defined to be effective only if participants achieve high awareness of their choices. Therefore, InViTo aims at providing a support for the communication and understanding of information, as key-elements for satisfying practitioners' requirements. In order to achieve this task, InViTo makes use of dynamic maps, which translates spatial and non-spatial data into a visual and intuitive language. Thus, the resulting maps can be used for exploring information, increasing the possibilities of understanding data related to complex systems such as cities and territories.

InViTo uses Geographic Information System (GIS) data in combination with different kinds of input, such as spreadsheets, vector and raster files. It is composed by two modules, one web platform for the production of multicriteria weighted maps [19, 20], and one visualisation module, which is based on a parametric modelling system [14].

Within the CODE24 project only the visualisation module has been used. In fact, it has been applied during collaborative workshops, in which the multicriteria analysis of strategic assessment has been carried out by means of the ANP

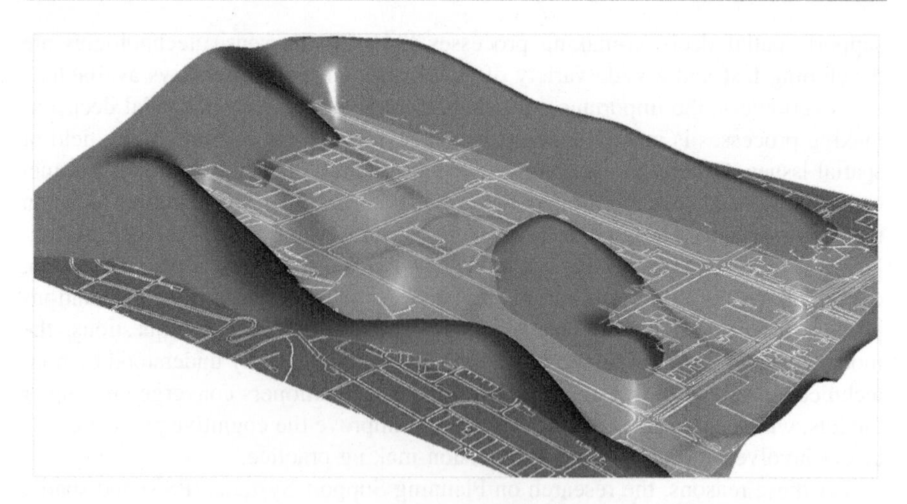

**Fig. 10.1** Example of dynamic map generated in Grasshopper by means of CAD and GIS files

evaluation technique [3, 22, 23]. As far the application of the method, please refer directly to the case studies (Chaps. 12 and 13).

The visualisation module of InViTo is based on Grasshopper, a free plug-in of Rhinoceros for the three-dimensional modelling. Since Grasshopper works on parametric features, the visualisation module of InViTo provides users with a tool for the generation of dynamic maps, which acts as multi-dimensional geo-referenced diagrams laying on different kind of cartographic support, such as CAD files or Google Maps viewer (Fig. 10.1).

In combination with the ANP assessment procedure, InViTo is set up to portray the relationships between a planning choice and the effect on a specific area. In particular, it provides a visual feedback which uses a cartographic background to localise where the consequences of planning choices may relapse. In addition, these dynamic maps show whether the effect is expected to be positive or negative using colours and its presumed intensity by means of diagram height.

In this way, InViTo can offer a common language to involve a wide range of users, who are generally characterised by different geographical and disciplinary backgrounds.

## 10.3 Visualisation as a Support to the Discussion on the Genoa-Rotterdam Corridor

In order to support the strategic assessment for the development of Corridor 24, visualisation has been used in different ways, with various purposes and at different spatial scales. As suggested by MacEachren et al. [10], geovisualisation can satisfy four main purposes: presentation, synthesis, analysis and exploration of data. Within the CODE24 project, each function of visualisation has been used to deal with specific issues of the planning topic, demonstrating how visual communication can provide high flexibility in supporting the different steps of planning and decision-making processes.

### 10.3.1 Data Presentation and Synthesis: The Video Production

The first use of visualisation within the CODE24 project has been the presentation of data to a large public in order to inform people about the project itself. The realisation of a video has provided the possibility to explain the concepts and strategies behind the project, as for instance the video animation which was used to illustrate the whole Corridor 24 Development Project during the Mobile Exhibition in Turin [11]. However, videos can also be an effective tool for informing people about an event, a project or the critical points of a problem. Thus, other videos have been made and shown at the beginning of collaborative workshops [12, 13] in order to present the particular issues characterising specific sections of the corridor. In this case, videos have widely contributed to inform participants about the main features of the specific area and have provided a rapid overview on the differences between development scenarios.

Although video animations do not offer any interactive features, they can be used to inform a large audience which is neither expert nor technician. This kind of data presentation works as a monologue of the video producer towards an undetermined public. Only selected elements are illustrated, and the final achievement is a general sharing of carefully chosen information. Videos preclude interaction and users can act only as mere spectators. However, videos are the visual product of a selection of filtered data, which are clustered in a defined framework and presented in a particular perspective. For this reason, videos can be conceived as a model, by which reality is summarised and proposed as an abstract synthesis. In this sense, they act as a means of transferring knowledge. They reproduce a vision of reality, albeit partial, and allow viewers to open their minds to a particular point of view. Nevertheless, as a form of visual language, videos provide the opportunity to quickly explain particular elements and simultaneously reach a large number of people.

### 10.3.2 Data analysis: The Construction of Maps

At the beginning of the CODE24 project, the main activity has been the gathering and analysis of data. Considering the vast size of the project's area, ranging from Rotterdam (NL) to Genoa (IT), the data collection has provided a huge amount of information which needed to be organised and structured in a readable framework. However, a collection of data is useful only if it can be analysed, but they are often incomplete with different systems for classifying records and assigning attributes. Therefore, in order to work with a large bulk of data, a methodology was required. To achieve this task, GIS tools provided effective support. In fact, they offer the

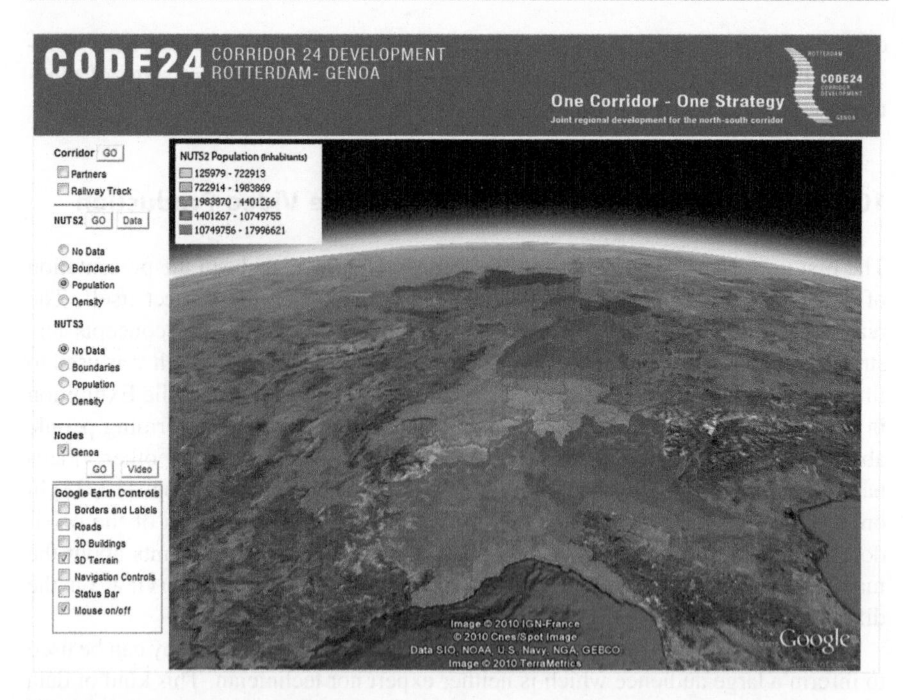


**Fig. 10.2** R&Set tool: visualisation of population diagram along the corridor 24 [16]

opportunity to organise georeferenced data within a framework of layers and tables. In particular, they associate spatial data with a position and a shape on a cartographic layout, thus providing a visual interface for reading the information. Therefore, the modality of displaying GIS data has been an essential example to understand how visualisation can support the analysis of spatial data. In fact, it can act as a scientific methodology for analysing data, because it provides a structure that allows data to be investigated, filtered and selected. Then, visualisation is a method for organising information within a framework and for analysing data, so that the reading of records can be facilitated by the intuitive process of visual reading. For the development of Corridor 24, a web page has been created in order to navigate through the data relative to the project area. It is a visualisation tool, namely the R&Set tool [16] (Fig. 10.2).

The R&Set tool considers the main data relative to the whole corridor, and it can be easily implemented by the partnership of the project. It displays spatial data within the virtual globe of Google Earth, exploiting the well-known user interface of this viewer. It provides a framework for the organisation of data, whatever they are, so to improve the users understanding of the information. In particular, it provides some possibilities of interaction with data, enabling users to have an active approach in developing their knowledge further. Therefore, this GIS visualisation tool can provide more information than a video, because it requires that users interact with the data model, investigating the displayed elements to their own purposes.

### 10.3.3 Data Exploration: The Knowledge Building

In combination with the project, visualisation has also evolved towards more sophisticated and effective uses. Despite the first part of the project when it was used to collect, organise and discuss data, the following steps of CODE24 project have been very intense in debating the strategic development of the corridor. During these steps, a number of experts have been grouped together to discuss the key-elements required to set up a complete common strategy. The experts were selected on the basis of their geographical origin and disciplinary skills, so to obtain a variegate mix of interests and knowledge. This mix is obviously a superb opportunity for granting a multifaceted overview on a future strategy. Nevertheless, it generates a lot of problems in communication: the communication among the people involved, but also the communication between the data model and user. Furthermore, such kind of collaborative processes require that the involved participants could achieve awareness of the elements affecting the planning in question, so to facilitate the optimal decisions as much as possible.

To overcome these problems, once again, visualisation can provide a valid support. It offers an effective methodology to deal with decision-making problems, giving the opportunity to use a visual language which overcomes national speeches and that provides the basis for the creation of a common table for enabling discussions. In addition, the interaction with data allows a dialogue between data and user to be built. In fact, literature agrees in recognising the combination of visualisation with high interaction as the basis for the exploration of data [10], which enables expert users to investigate those hidden connections [5] that relate the different elements to each other.

Therefore, to support these collaborative assessment workshops, the ANP procedure has been flanked by InViTo, which provided both the interactive and visual structure for carrying out the discussion (Fig. 10.3).

During the 4 year long project, the fast development of information technologies brought many novelties within the field of geovisualisation tools. While instruments related to geographic studies had a very fast growth, the widespread use of Google Earth and the outbreak of social networks conveyed the research on geographic information towards more dynamic systems of visualisation. In the same way, during the CODE24 project, the spatial data visualisation has evolved towards more effective forms of representation. From simple data maps built in a Google Earth environment, the introduction of a parametric modelling tool within the same virtual globe brought a new dynamic representation of geo-related data. The interactive visualisations provided by InViTo have offered new opportunities to the use of GIS data within collaborative processes. Each geometry within the model can now be related to various typologies of data, both spatial and non-spatial, so to connect georeferenced shapes with multi-variegated sets of information. Thus, decision-making issues can be associated to a spatial effect on the area, showing the consequences of particular choices.

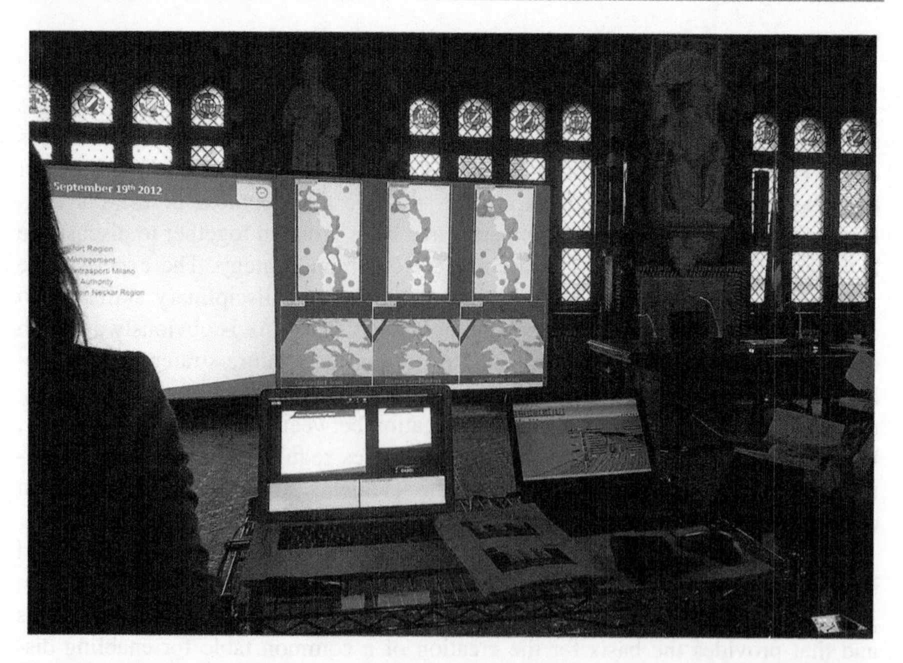


**Fig. 10.3** The use of InViTo during an ANP workshop (september 2012, at Genoa Port Authority, Genoa, Italy)

In particular, the use of InViTo allowed information to be explored. During the discussions, maps were built on the basis of participants' opinions in real time. People involved in the workshop could see the "what if" representations of their choices, allowing them to evaluate their answers. This provided an effective way for understanding how a choice can positively or negatively affect the overall strategy on the whole corridor.

## 10.4 Overview on the Use of Geo-Visualisation Within the CODE24 Project

The application of interactive visualisation to the ANP assessment technique generated several questions in relation to the used language, the users' perception, the ease of understanding maps and the maps' ability of reproducing the ANP results. The CODE24 project provided the opportunity to study the use of geo-visualisation within actual multi-disciplinary collaborative sessions. Table 10.1 illustrates the use of geovisualisation within the project for the development of the Genoa-Rotterdam corridor, showing its progress during the 4 years.

Table 10.1 highlights a process of growth, which started in 2010 with the visualisation of simple 3D data maps and ended in 2013 with interactive charts, enveloping different purposes and spatial scales.

**Table 10.1** Development of visualisation during the 4 years project

| Time | Image | Typology of visualisation | ANP network | Function of visualisation | Strengths and weaknesses |
|---|---|---|---|---|---|
| Nov. 2010 | | GIS data + Google earth (R&Set tool) | No ANP | Analysis | **S**: easiness in reading GIS data<br>**W**: no contribution by project partners |
| Jan. 2011 | | 3D models + GIS data + Google earth (R&Set tool) | No ANP | Analysis | **S**: hyper-information<br>**W**: no changes in forms and functions |
| Avr. 2011 | | InViTo, metropolitan 3D dynamic maps | BOCR network | Exploration | **S**: localisation of the ANP issues<br>**W**: confusion during the overlapping of all the maps |

(continued)

**Table 10.1** (continued)

| Time | Image | Typology of visualisation | ANP network | Function of visualisation | Strengths and weaknesses |
|---|---|---|---|---|---|
| Sept. 2011 |  | Video animation | No ANP | Presentation | **S**: eye catcher<br>**W**: monologue |
| July 2011 | 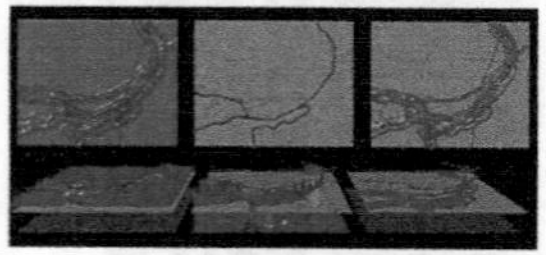 | InViTo, sub-regional 3D dynamic maps | BOCR network | Exploration | **S**: localisation of the ANP issues<br>**W**: too many maps and information |
| Dec. 2011 | 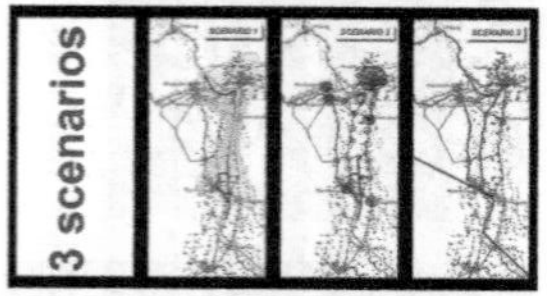 | Video animation | No ANP | Presentation | **S**: quick topic description<br>**W**: monologue |
| Dec. 2011 | 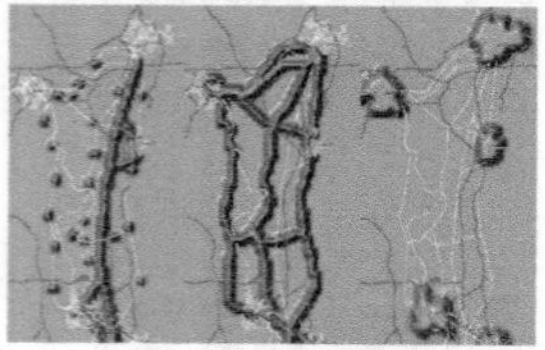 | InViTo, regional 3D dynamic maps | BOCR network | Exploration | **S**: localisation of the ANP issues<br>**W**: maps symbolise the node and not the effect of the node |

(continued)

**Table 10.1** (continued)

| Time | Image | Typology of visualisation | ANP network | Function of visualisation | Strengths and weaknesses |
|---|---|---|---|---|---|
| Mar. 2012 |  | Video animation | No ANP | Presentation | **S**: quick topic description<br>**W**: monologue |
| Mar. 2012 | 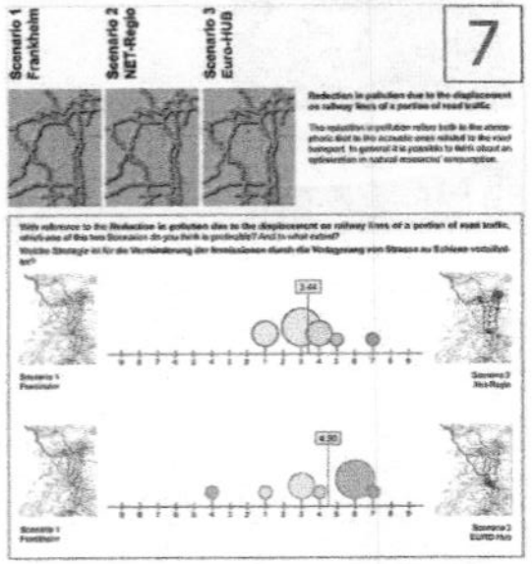 | Bubble charts | BC network survey answers | Analysis | **S**: readability of answers<br>**W**:difficulties in comparing with other bubble charts |
| Mar. 2012 | 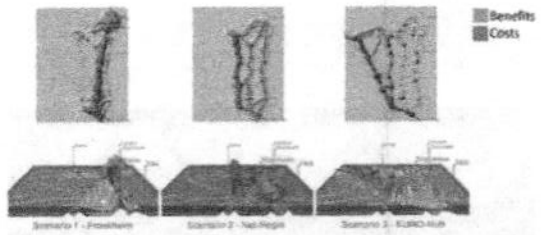 | InViTo, regional 3D dynamic maps | Complex Strategic-BC network | Exploration | **S**: localisation of the expected effects of ANP nodes<br>**W**: high symbolism which must be previously explained |

(continued)

**Table 10.1** (continued)

| Time | Image | Typology of visualisation | ANP network | Function of visualisation | Strengths and weaknesses |
|---|---|---|---|---|---|
| Sept. 2012 | 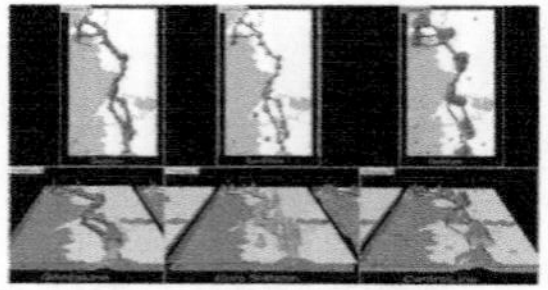 | InViTo, transnational 3D dynamic maps | BC network | Exploration | **S**: localisation of expected benefits and costs |
| | | | | | **W**: large size |
| Oct.-Dec. 2012 | 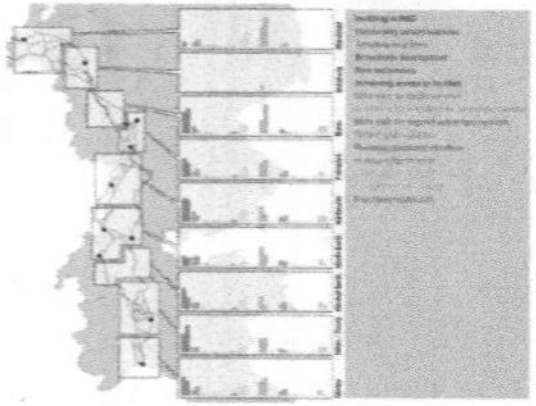 | InViTo, 2D geo-related histograms | Simple ANP network | exploration | **S**: detailed information for each sub-region |
| | | | | | **W**: too many charts. |
| Jan. 2013 | 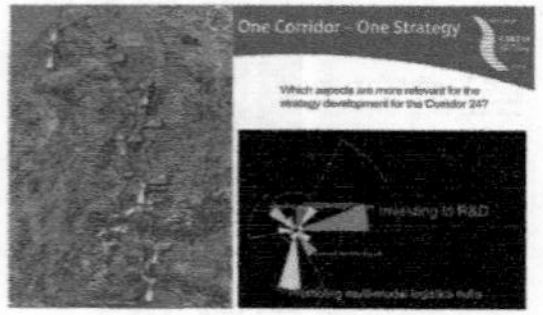  | InViTo, 2D geo-related radial charts | Simple ANP network | exploration | **S**: information on the effects on each sub-region. |
| | | | | | **W**: lack of data |

(continued)

**Table 10.1** (continued)

| Time | Image | Typology of visualisation | ANP network | Function of visualisation | Strengths and weaknesses |
|---|---|---|---|---|---|
| Feb. 2013 | 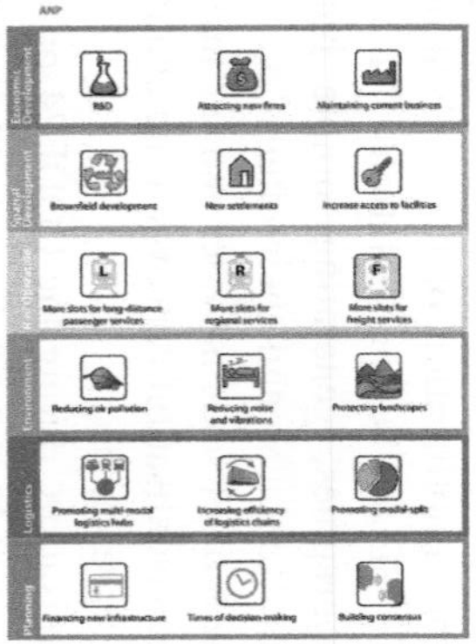 ANP<br>Economic Development: R&D; Attracting new firms; Maintaining current business<br>Spatial Development: Brownfield development; New settlements; Increase access to facilities<br>More slots for long-distance passenger services; More slots for regional services; More slots for freight services<br>Environment: Reducing air pollution; Reducing noise and vibrations; Protecting landscapes<br>Logistics: Promoting multi-modal logistics hubs; Increasing efficiency of logistics chains; Promoting modal split<br>Planning: Financing new infrastructure; Times of decision-making; Building consensus | Icons | Simple ANP network | analysis | **S**: intuitiveness<br>**W**: too numerous |
| Feb. 2013 | 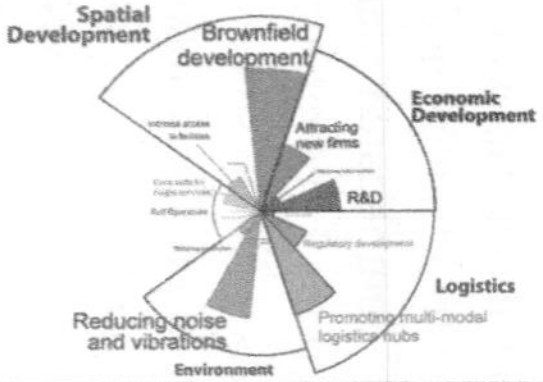 Spatial Development<br>Brownfield development<br>Economic Development<br>Attracting new firms<br>R&D<br>Logistics<br>Promoting multi-modal logistics hubs<br>Reducing noise and vibrations<br>Environment | InViTo, a single interactive 2D radial chart. | Simple ANP network | exploration | **S**: intuitive and concise<br>**W**: no information on spatial consequences |

The CODE24 project has been carried out by studying different sections of the corridor Genoa-Rotterdam, concluding with the development of a strategy for the whole corridor. For each step, a set of workshops have been carried out by combining the use of the ANP assessment technique and InViTo. Therefore, each workshop has been supported by an interactive visualisation in order to enable a visual understanding of ANP issues and the exploration of information within the maps (Chap. 6, Lami). The first workshop in April 2011, made use of the whole ANP structure, that is the BOCR network, where BOCR means Benefits, Opportunities, Costs and Risks. In the first study, InViTo has been set up to produce four dynamic maps for each strategy option, and a map for each BOCR network. This structure was designed for a metropolitan scale on the Wesel area and was re-proposed several months later, to a larger scale, to analyse the Bellinzona case study (Chap. 12, Abastante et al.). For both cases, the BOCR maps have provided very interesting, but overly dense information. The following workshops for the assessment of the region between Frankfurt and Mannheim brought a simplification to the ANP framework, and changed it from a BOCR to a BC network Chap. 7, Lami. This lighter framework decreased the total amount of maps and provided another important change in the visualisation. During the first uses, maps were built symbolically using the spatial shape of a geographical or an infrastructural element. With the introduction of a simpler network, maps were fewer in number and split between positive and negative effects. This made it possible to build maps which localise where each ANP element is expected to have a spatial effect.. This change brought spatialisation into the discussion, raising the debate on well localised spatial systems.

Nevertheless, during the study on the Frankfurt-Mannheim case, the spatialisation showed to be a very important element for supporting the discussion, the passage to a trans-national scale revolutionised the requirements for supporting the ANP discussion. In fact, when facing a very wide area, the development strategy is generally considered as a common goal, so that local disadvantages should lose their importance with respect to a common advantage. The same happened in visualising the strategy: the localisation of costs and benefits was not as important as the typology of the effect and its intensity. Therefore, during the workshops concerning the whole corridor of Genoa-Rotterdam, the visualisation lost its relationship with the geo-location of data and became a way for analysing different single strategies. In the same way, the simplification of the ANP structure to a simple network, led to a transition from 3D to 2D representations. Thus, 2D radial charts substituted complicated cartographies, generating simple but easy-to-understand connections between actors' choices and their effects on the overall corridor strategy. As a result, since the project scale is wider and the need for geo-relating the effects of choice is lower, visualisation can overcome spatial barriers and move to more abstract forms of representation.

In addition to the interactive visualisation, during the CODE24 project, other typologies of visualisation have been used to support the information process of the people involved. Video animations proved to be very effective in focussing the participants on the planning issue. Their use during the presentation of the workshop has been essential to quickly summarize the key-features of the meeting and illustrate the main strategies to be discussed. A further support of visualisation has been given by info-graphics: bubble charts and icons have provided a methodology for rapidly conveying information, both quantitative (by charts) and qualitative (by icons).

Workshop by workshop, visualisation has evolved in order to better deal with the questions which emerged during the debates, highlighting the need for simplicity in structuring both the data model and the communication. This experience has proven that visualisation needs to be designed and cannot just be the fruit of an aesthetical composition of shapes and colours. The visualisation is the framework through which data has been communicated to the participants during workshops. It deeply determines how information is acquired and modified in the choices of each actor.

## 10.5 Conclusions

Visualisation has demonstrated to be a valid support tool for all the steps of a wide area project such as the Rotterdam-Genoa corridor. It has been used with different purposes and users, to create a common ground for the conversation and discussion of both technical and non-technical issues.

Visualisation has been adapted in different tools to translate data into knowledge by means of different levels of interaction. At the lower level, data has been presented to large audiences and videos have proven to be an effective media tool. At a medium level, data has been structured in a defined framework which can be defined as a data model. which This data model can be used for analysing data through the collection, filtering, selection and organisation of data. Finally, at the higher level, the data exploration uses the data model to understand the inner relationships among data.

Therefore, interactive visualisation has brought new practices for dealing with decision-making processes. It has enabled new cognitive methodologies and has enhanced the collaboration and awareness of every single actor.

## References

1. Andrienko G, Andrienko N, Jankowski P, Keim D, Kraak MJ, MacEachren AM, Wrobel S (2007) Geovisual analytics for spatial decision support: Setting the research agenda. Int J Geogr Inf Sci 21(8):839–857
2. Andrienko G, Andrienko N, Keim D, MacEachren A, Wrobel S (2011) Challenging problems of geospatial visual analytics. J Visual Lang Comput 22(4):251–256
3. Bottero M, Lami IM, Lombardi P (2008) Analytic network process. La valutazione di scenari di trasformazione urbana e territoriale, Alinea, Firenze
4. Couclelis H (2005) Where has the future gone? Rethinking the role of integrated land-use models in spatial planning. Environ Plan A 37(8):1353–1371

5. Dodge M Information maps: tools for document exploration. CASA, working paper series 94, (July 2005)
6. Heer J, Agrawala M (2008) Design considerations for collaborative visual analytics. Inf Vis 7(1):49–62
7. Klosterman RE (2012) Simple and complex models. Environ Plan 39(1):1–6
8. Klosterman RE, Pettit C (2005) Editorial: an update on planning support systems. Environ Plan 32:477–484
9. MacEachren AM, Cai G, Hardisty F (2003) Geovisualization for knowledge construction and decision-support. In: T.P. University (eds) Retrieved Mar 20, 2013, from GeoVISTA Center. http://www.geovista.psu.edu/publications/2003/MacEachren_CG&A_03.pdf
10. MacEachren AM, Gahegan M, Pike W, Brewer I, Cai G, Lengerich E, Hardisty F (2004) Geovisualization for knowledge construction and decision-support. Comput Graph Appl 24(1):13–17
11. Masala E (2011) CoDe24—MobileExhibition—sept 2011—Torino. From YouTube. http://youtu.be/KPwGCId10fo
12. Masala E (2012a) ETH Dec 13, 2011—Computational Assessment Workshop. Retrieved June 8, 2012, from YouTube. http://youtu.be/PSnu_Ti66VU
13. Masala E (2012b) ETH Mar 20, 2012—computational assessment workshop. Retrieved June 8, 2012, from YouTube. http://youtu.be/6ghuI0h1YRU
14. Masala E, Marina O, Pensa S, Stavric M (2012) Interactive visualization in modeling urban development. In: Billen R, Caglioni M, Marina O, Rabino G, San José R (eds) cost action TU0801—3D issues in urban and environmental systems (pp 59–65). Escolapio, Milano
15. McCormick BH, De Fanti TA, Brown MD (1987) Visualization in scientific computing. Comput Graph 21(6):69
16. Pensa S (2011) R & Set corridor visualization tool. Retrived Mar 15, 2012 from. http://v-tool.code-24.eu/code24.html
17. Pensa S, Masala E (2014a) The interactive visualisation tool (InViTo): concepts and usability. In Masala E, Melis G (eds) interactive visualisation tool for brownfield redevelopment—a european experience (*Forthcoming*). Celid, Torino
18. Pensa S, Masala E (2014b) InViTo: an interactive visualisation tool to support spatial decision processes. In: Pinto NN (eds) virtual cities and territories (*Forthcoming*). IGI Global Book
19. Pensa S, Gagliarducci R, Masala E (2013) Retrieved from InViTo—interactive visualization tool. http://invito.urbanbox.it/
20. Pensa S, Masala E, Lami IM (2013) Supporting planning processes by the use of dynamic visualisation. In: Geertman S, Toppen F, Stillwell J (eds) Planning support systems for sustainable urban development, vol 195. Springer, Berlin, pp 451–467
21. Pensa S, Masala E, Lami IM, Rosa A (2014) Seeing is knowing: data exploration as a support to planning. Civ Eng Spec Issue 167(CE5), pp 3–8
22. Saaty TL (1980) The analytic hierarchy process. McGraw Hill, New York
23. Saaty TL, Vargas LG (2006) Decision making with the analytic network process. Springer, New York
24. te Brömmelstroet MC (2010) Equip the warrior instead of manning the equipment: land use and transport planning support in the Netherlands. J Transp Land Use 3:25–41
25. Thomas JJ, Cook KA (2005) Illuminating the path: the R&D agenda for visual analytics. IEEE Press
26. Tufte ER (1990) Envisioning information. Graphics Press, Cheshire
27. Tufte ER (2001) The visual display of quantitative information. Graphics Press, Cheshire
28. Vonk G, Geertman S, Schot P (2005) Bottlenecks blocking widespread usage of planning support systems. Environ Plan A 37(5):909–924

# Part IV
# Case Studies

# An Integrated Approach for Supporting the Evaluation of Transport Scenarios: The Area of Bellinzona (CH)

11

Francesca Abastante, Marta Bottero, Felix Günther, Isabella M. Lami, Elena Masala, Stefano Pensa and Ilaria Tosoni

Abstract

Multiple Criteria Decision Analysis (MCDA) is a widely-used tool to support decision processes when a choice between different options is needed. This approach is particularly useful in situation characterized by an inter-connected range of environmental, social and economic issues. Moreover the presence of many actors with different backgrounds and knowledge constitutes a further

F. Abastante (✉) · M. Bottero · I. M. Lami
Department of Regional and Urban Studies and Planning (DIST),
Politecnico di Torino, Corso Massimo D'Azeglio 42, 10123 Turin, Italy
e-mail: francesca.abastante@polito.it

M. Bottero
e-mail: marta.bottero@polito.it

I. M. Lami
e-mail: isabella.lami@polito.it

F. Günther · I. Tosoni
Institute for Spatial and Landscape Development Chair for Spatial Development,
ETH Zürich, Wolfgang-Pauli-Str. 15 (HIL H 37.4), 8093 Zürich, Switzerland
e-mail: fguenther@ethz.ch

I. Tosoni
e-mail: ilaria.tosoni@gmail.it

E. Masala · S. Pensa
SiTI-Higher Institute on Territorial Systems for Innovation,
via Pier Carlo Boggio 61, 10138 Turin, Italy
e-mail: elena.masala@polito.it

S. Pensa
e-mail: stefano.pensa@polito.it

I. M. Lami (ed.), *Analytical Decision-Making Methods for Evaluating Sustainable Transport in European Corridors*, Sxi 11, DOI: 10.1007/978-3-319-04786-7_11,

level of complexity due to the difficulty in interpreting and reading outputs. The present application is one response to tackle these difficulties. It is an innovative approach integrating Analytic Network Process (ANP) and Interactive Visualization Tool (InViTo) which creates a common language among the actors involved and a shared basis for generating discussion. The methodological framework is applied to a Swiss section of the transport corridor Genoa—Rotterdam, within the Interreg IVB NWE Project "Code24" in order to demonstrate the potential of the joint use of the two tools mentioned for the selection of a suitable strategy for transport improvement within territorial transformation.

## 11.1 Introduction

The recent evolution of the decision making process related to the implementation of European corridors has stimulated a broad debate not only on the integration of transport and land use, but also on the need for new tools to support the decisions made in that specific field.

The underlying assumption is that there is no pre-established appropriate strategy. The open issue is how the new infrastructure corridors will benefit from their operational and technological autonomy and how, conversely, they have to interact with local history and specificity in order to positively affect every regional area as a whole. Every regional and urban node must be interpreted in relation to different levels of complementarity and polycentric integration: it is the transition from a society of places to a society of flows that requires the consideration of multiple levels of relationship, scale and intensity [35].

The need for decision support tools, which are able to simultaneously consider different aspects of transport planning, is becoming increasingly evident.

In this complex context, it has been generally agreed that Multiple Criteria Decision Analysis (MCDA) can provide a very useful support. MCDA has been used to make comparative assessments of alternative projects or heterogeneous measures [9, 25]. These methods allow several criteria to be taken into account simultaneously in a complex situation. They are designed to help Decision Makers (DMs) to integrate the different options, which reflect the opinions of the actors involved, in a prospective or retrospective framework.

Although MCDA is widely used to support decision making/aiding processes, difficulties in reading output data can often limit the process of data and knowledge sharing. One important difficulty is the diversity of the DM's backgrounds. This research approaches the integration between MCDA and data visualization in order to create a shared basis among the actors involved in the decision process [4, 14, 15]. The use of an interactive visualization tool can support MCDA in showing results, exploring alternative options and evaluating the differences in the localization of the expected positive and negative effects. Therefore, a methodological framework involving Analytic Network Process (ANP) [29, 31] and the Interactive Visualization Tool (InViTo) has been studied and applied [20, 35].

This chapter presents an application to a Swiss section of the Corridor 24, part of a Genoa-Rotterdam, Interreg IVB NWE Project, called "Code24". The Code24 Project aims at identifying a shared spatial and infrastructural development strategy for the regions connected through this infrastructure of strategic European importance. In order to come to a shared strategy for the corridor, it is important to come to a common understanding about the unsolved issues affecting the different regions. This means, first of all, to survey the consequences of the pending decisions regarding alternative strategies and interventions. In order to achieve this result an assessment procedure has been developed in the project's framework to accompany the discussion in the areas where priorities and development strategies are not yet clear and need to be set. The assessment procedure has been applied in several international workshops organized along the Corridor, in order to jointly review with local actors and stakeholders in the different areas.

The aim of the application of the assessment procedures illustrated herein concerns the classification of three development scenarios regarding the area of Bellinzona. After the introduction, the chapter is organized as follows: Sect. 11.2 illustrates the methodological framework; Sect. 11.3 describes the structure of the ANP/InViTo model and shows the results of the application; finally, Sect. 11.4 concludes the chapter by highlighting the strengths and weaknesses of the proposed joint methodological framework.

## 11.2 Methodological Framework

### 11.2.1 The Analytic Network Process

The Analytic Network Process (ANP) is a multi-criteria methodology which is able to consider a wide range of quantitative and qualitative criteria, according to a complex model [27, 28]. It structures the decision problem into a network and uses a system of pairwise comparisons to measure the weights of the structure components and to rank the alternatives. The ANP model consists of control hierarchies, clusters and elements, as well as interrelations between elements because it is able to connect clusters and elements in any manner in order to obtain priority scales from the distribution of the influence between the elements and clusters. The structure of the model is characterized by continuous feedback between the elements and the cluster and it is able to capture the complexity of the reality [30].

The application process of the ANP can be summarized in four main phases:

Step 1: *Structuring the decision problem and model construction.*
There are two types of models that can be developed within the ANP methodology: the complex network model and the simple structured model [28, 32].

Step 2: *Compilation of pairwise comparison matrices.*
Having constructed the decision model and having established relations between the elements, it is possible to proceed with the pairwise comparisons between the elements. The evaluation takes place in two levels: that of the clusters and that of the nodes. In pairwise comparisons, a ratio scale of 1–9, that is Saaty's fundamental scale, is used to compare any two elements where value 1 indicates that the two elements are equally important and value 9 indicates that the difference between the two decision elements is extremely significant. The assigned ratings are placed in a matrix of pairwise comparisons [26]. The main eigenvector of each pairwise comparison matrix represents the synthesis of the numerical judgements established at each level of the network [26].

Step 3: *Construction of supermatrices.*
A supermatrix represents, in the case of the ANP, the relationships that exist within the network model and the relative assigned weights. It is an array containing all the priority vectors that are extracted from individual pairwise comparison matrices compiled during the previous steps of analysis. Firstly, the supermatrix plays a fundamental role in the analysis because it allows an understanding of certain relationships of influence determined during the development of the network. Secondly, the supermatrix is crucial because, being composed by different eigenvectors, it provides numerical data about the priorities of elements forming part of the decision system [5]. During the development of the ANP methodology, three different supermatrices are extracted:

- The unweighed supermatrix (or initial supermatrix), which contains all the eigenvectors that are derived from the pairwise comparison matrixes of the model.
- The weighted supermatrix, which is a stochastic supermatrix obtained by multiplying the values in unweighed supermatrix by the weight of each cluster. In this way it is possible to consider the priority level assigned to each cluster.
- The limit supermatrix, which is the final matrix of the analysis obtained by raising to a limiting power the weighted supermatrix in order to converge and to obtain a long-term stable set of weights that represents the final priority vector.

Step 4: *Elicitation of the final priorities and sensitivity analysis.*
In the case of the complex network structure, it is necessary to synthesize the outcome of the alternative priorities for each of the different subnetworks in order to obtain their overall synthesis through the application of different aggregation formulas [30]. The last step consists in carrying out the sensitivity analysis on the final outcome of the model in order to test its robustness.

The literature is quite recent and some publications can be found in strategic policy planning [40], market and logistics [3], economics and finance [17] and in civil engineering [18, 23], in territorial, transport and environmental assessment [1, 2, 5–7, 13, 24, 38].

### 11.2.2 Visual Maps

In spatial planning, visualisation refers to the exploration of spatial data, which is a discipline known as geovisualisation. Dynamic maps and georeferenced charts are the visual methods which are becoming ever more effective for communicating such information, offering real time responses to users' queries and showing data, data clusters and relations among data. By using geovisualisation tools, spatial decision processes can significantly benefit from informing the actors involved, enhancing discussions and creating awareness of choices to be taken.

Geo-visualization is a branch of cartographic science and is defined as a technique for the exploration of spatial and spatio-temporal data through the use of interactive tools [4]. Literature on geovisualization provides different examples of three-dimensional models coming from GIS data [16]. These representations are mainly based on generative modelling [41], which results in the automatic production of three-dimensional volumes directly from databases, model libraries or through the extrusion of specific database attributes. Many tools provide these types of spatial data visualization such as "Community Viz" (Orton Family Foundation and Placeways, LLC) and "Metroquest" (Envision Sustainability Tools Inc.), but use it primarily for project presentation, instead of data exploration during the planning process. There is a lack of systems able to integrate the generation of 3D volumes and tools which make use of parametric functions.

To address these issues, a new approach to decision making has been developed, which could provide an effective framework for the construction of discussion and knowledge. This research resulted in the Interactive Visualisation Tool (InViTo), a visual method for managing spatial data in real time (see Chap. 10).

InViTo aims at combining the elements of spatial problems with their corresponding geographical effects by using dynamic maps which change according to DMs' choices. To achieve this task, InViTo organizes data and the relationships between them in a visual interface, which allow DMs to analyse and explore spatial objects in real-time. InViTo is a tool conceived as a Planning Support System (PSS) and Spatial Decision Support System (SDSS) for aiding the actors involved in sharing information and raising awareness of spatial issues at different scales [11, 12, 19, 21, 22];

In order to combine the ANP and the Invito methodologies, each BOCR element is associated with a map, which receives a weight by the actors involved in the evaluation process. The model of visualization built in InViTo sums up all the selected maps in a singular 3D mesh. This new geometry defines the intensity of the positive or negative effects on the area by means of peaks, whose heights depend on the values of the weights given by the actors and on the chosen scenario. Secondly, users can interact with the 3D model. The visualization system allows the values of the weight to be input in real time and, at the same time, it displays the changes which occur on the relative maps in real time so that the different decisions made by the actors can be readily visualized and, then, compared.

Thanks to this ability to interact with users' proposals, the visualization of data offers a methodology for explaining the relationships of cause and effect occurring between actors' decisions and spatial configuration. In fact, these maps are used to help the actors involved in understanding two main issues:

1. The correlation between their decision and territory.
2. How the assignment of a value to the weight between the different elements can vary the result.

## 11.3 Presentation of the Case and Illustration of the Alternatives

There are two agreements between Switzerland and the European Union (EU) which allow the cross-border land transport with no limitations despite the fact that Switzerland is not one of the Countries party to the EU (CEE/Switzerland, 1992 and Switzerland/CE 1999). Furthermore Switzerland has the obligation to transfer goods from border to border and from road to rail (Constitution art. 36/84 Alpenquerender Transitverkehr initiative 1994/99; delocalization ACT 1999).

For these reasons Switzerland would fit into the nascent European rail network in the best possible way.

To reach this objective, significant projects have been developed in Switzerland under the name "AlpTransit projects" (NEAT) (Fig. 11.1).

The key elements of the NEAT are the Gotthard Tunnel and the Löschtberg Tunnel together with the construction of new access roads. A fundamental project inserted in the NEAT system is the axis that connects Zurich (Switzerland) and Milan (Italy) and more specifically the Gotthard–Lugano portion for which the locations of stations and railway tracks have not yet been decided [11].

Economic and technical agreements between the countries involved (i.e. Germany, Switzerland and Italy) are needed in order to continue the planning of the railway layout. In fact, this has led to considerable delays of several important projects in the north–south axis: the Gotthard Tunnel is expected by 2016/2017 while the Zimmerberg Tunnel, the Hirzler Tunnel and some fundamental roads have been indefinitely postponed. In 2010 a new plan called "STEP" identified new priorities until 2025 considering the completion for the works in progress but at the same time limiting the financial resources destined for new infrastructures. STEP was developed in response to the demand for transport, which is higher on the east–west axis but lower on the north–south axis due to the on-going process of depopulation.

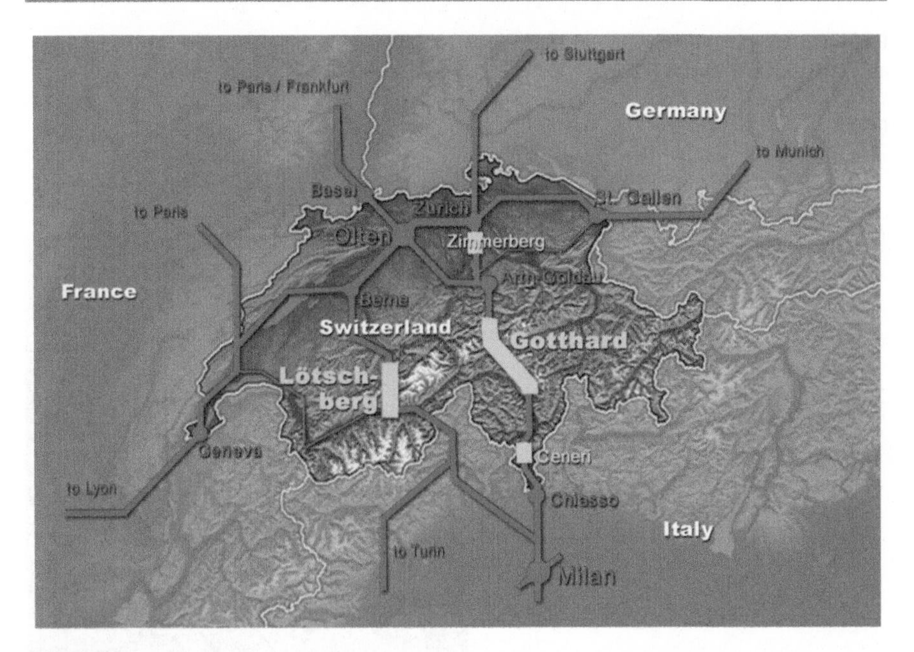


**Fig. 11.1** The AlpTransit project (*Source* ETH Zurich IRL for INTERREG IVB NWE project CODE24, 2011)

Therefore the north–south axis is a crucial project that may improve or worsen the transport services depending on the operating conditions.

Spatial and territorial development is influenced by the expansion and the construction of railway infrastructure in terms of accessibility and presence of services to the population. These can enhance the attraction of an area from an economic, logistic, industrial and tourist point of view. Changes in accessibility conditions inevitably bring advantages and disadvantages to the regions affected. Therefore, accessibility is a crucial point for the territorial development on this spatial axis. Reducing travel time between Zurich and Milan means creating new daily commuter movements. This would cause major positive changes in the spatial and functional structure of the territories and in the way of living and working for residential people.

In this complex panorama, the territory of Bellinzona is in a particular situation: it will be the only site along the corridor track that will see a reduction in the travel time, starting from 2020.

Improving the accessibility of the area will open up new opportunities for a better use of brownfield sites around the train stations. Thus, the brownfield sites are strategic areas for the future development of Bellinzona. Furthermore, they constitute another challenge to the construction of the NEAT system: exploiting the potentiality of the new railway line while repositioning the workforce of the machine workshops. The attraction of a mobile workforce and of advanced

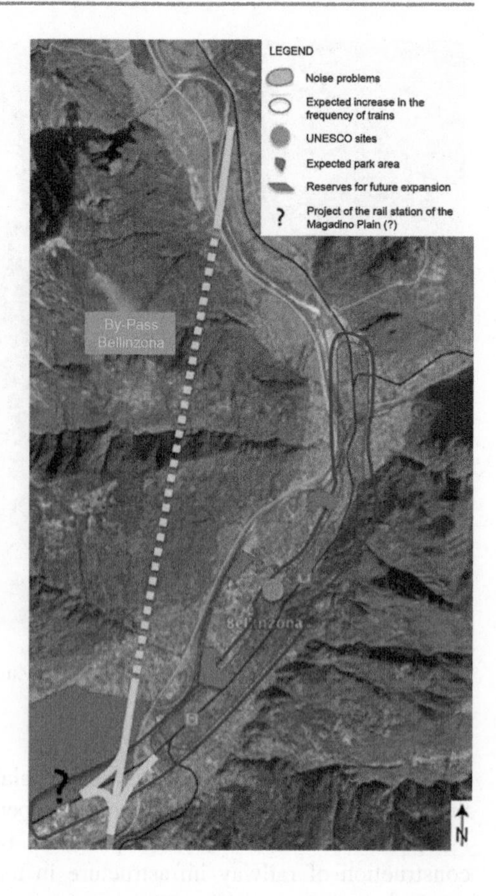


**Fig. 11.2** The Bellinzona area (*Source* ETH Zurich IRL for INTERREG IVB NWE project CODE24, 2011)

services in addition to the presence of specialized labour are fundamental for the economic success of this area.

The railway by-pass of the Bellinzona area is another issue that could affect the territory. The by-pass should convey the freight and transit traffic into a tunnel located in the west side of the settlement. It constitutes a strategic project that should solve the problems related to the noise pollution that affect the area. The risks are that the tunnel might also be used for fast passenger connections, causing a downgrade of Bellinzona into a regional railway node without solving the noise issues (Fig. 11.2).

Moreover, a further difficulty is constituted by the Magadino Plain. This is a flat connection area between Bellinzona and Locarno and it is the principal topic of discussion of the Ticino planning area. The Magadino Plain is an environmentally protected area where strict building constraints have been established in order to protect the environmental balance. Despite the willingness of the planners to preserve this natural area, the landscape has inevitably been affected by the real estate market (i.e. large shopping centres and luxury villas on the slopes of the surrounding mountains). The new infrastructural projects will reduce travel time

**Table 11.1** Alternative development strategies

| Alternatives | Description |
|---|---|
| Scenario 1 | The scenario involves the creation of a rail bypass which is destined both for passengers and freight. It implies the construction of a new rail tunnel as well as a new station in the Magadino area |
| Scenario 2 | The scenario concerns a series of works aiming at improving the existing rail line, in order to increase the transport capacity. Some mitigations of acoustic impact will be provided within this scenario |
| Scenario 3 | The scenario considers the construction of a rail bypass which will be exclusively destined to freight transport |

and will offer new opportunities of growth but it will also bring new development pressure to the Magadino Plain.

The Italian authorities are sponsoring the construction of a costly new freight corridor for the railway line of Novara (Italy) which will cross the Magadino Plain. This would bring important advantages for the Italian freight transport but it would produce devastating effects for the Magadino Plain.

Three alternative development strategies have been designed in order to sum up the current debate on the decision process for the Bellinzona area (Table 11.1).

## 11.4 Structuring of the Decision Problem

### 11.4.1 Definition of the BOCR Model

A complex ANP model has been developed in order to identify the best alternative development strategy for the area of Bellinzona. According to the literature [28], the decision problem has been divided into four clusters (namely environmental aspects, urban aspects, transport aspects and economic aspects) in turn divided into elements. Clusters and elements were organized according to a BOCR model (i.e. Benefits, Costs, Opportunities and Risk). In this model, Benefits and Costs have been considered, respectively as positive and negative aspects of the transformation at the present time, while Opportunities and Risks have been considered as positive and negative aspects of the transformation in future scenarios. Figure 11.3 shows the complete network of the model.

The choice of applying a complex BOCR network is related to the complex nature of the examined decision problem. In fact, it allows a high number of aspects occurring in different time periods to be taken into account. Table 11.2 represents the decision network of the problem.

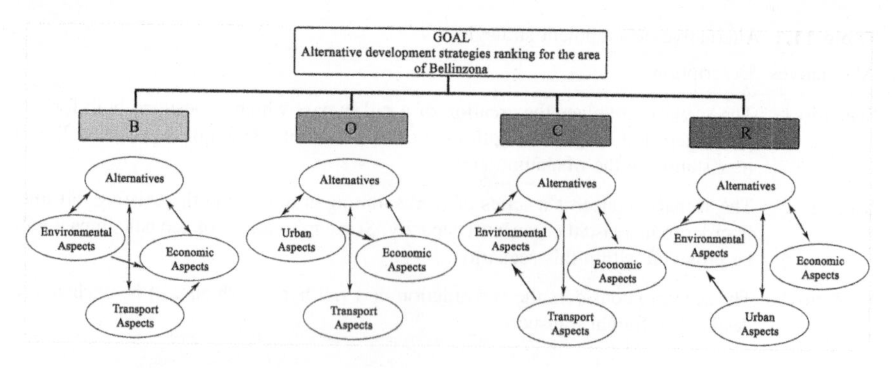


**Fig. 11.3** BOCR model

## 11.4.2 Construction of the Partial Maps

Since the ANP technique works by assigning weights to the relationships occurring among elements, clusters and subnets, InViTo has been set up in order to generate spatial forms depending on weighted relationships among the 3D model components. Therefore, the same ANP organization of relationships and weights has been reproduced to generate the 3D model. In this sense, each element considered by the BOCR model has been associated onto a spatial map which symbolically localises its effect on the area (see Table 11.3). For example, the map of Benefits visualising the increase of accessibility uses a buffer area around the railway stations, because this is the area expected to receive the most effects related to railway accessibility. Nevertheless, not all the elements have a spatial consequence which can be positioned in a specific place. For example, the positive effects regarding the expected increase of the attractiveness to the area (CT, Table 11.3), or the possible extensions of the implementation time derived from the conflicts among the local population, (IT, Table 11.3) are both elements which have no defined boundaries and can be difficult to draft on a map only with difficulty. Furthermore, they have a diffuse consequence on the whole region considered. Therefore, their visualization has been designed as a constant value covering the whole area. On the contrary, the elements with a localized effect on defined areas have been closely associated onto maps which represent diversified values depending on the localization of their effect. Such elements include the environmental and transport aspects which generally have a recognizable area of impact.

The peculiarity of these maps is represented by the possibilities given by the interaction with them. The actors involved in the process are able to see the changes that their evaluations produce, as well as, being able to locate the elements under discussion. In this way, decision-makers are assisted in confirming or changing their point of view.

**Table 11.2** Cluster and nodes of the BOCR model

| Clusters | Elements | | Description |
|---|---|---|---|
| *Benefits* | | | |
| Economic aspects | Valorization of the real estate market | RE | Valorisation of the real estate market in the area related to the improved accessibility |
| | Valorization of touristic local system | TL | Valorisation of the touristic local system in the area related to the improved accessibility |
| Environmental aspects | Conservation of protected area of Magadino | PA | Landscape protection and enhancement of territorial peculiarities of the area of Magadino |
| | Reduction in acoustic emission | RA | Reduction in acustic emission due to the use of compensation measures as acoustic barriers along the railway lines |
| Transport aspects | Increase in accessibility | IA | Increased accessibility due to the development of transport connections |
| | Increase in capacity of freight transport | CF | Optimization and development of freight transport connections |
| *Opportunities* | | | |
| Economic aspects | Creation of employments | CE | Creation of new employments directly related to the transport improvement |
| Transport aspects | Increase in connections between Ticino and Lombardy region | DT | Increase in transport connections between the Ticino area and the Lombardy region |
| Urban aspects | Creation of new urban centrality | UC | Creation of new urban centralities in the areas affected by the transformation |
| | Development of the Ticino area and of its attractiveness | CT | Increased attractiveness due to the development of transport connections |
| *Costs* | | | |
| Economic aspects | Acquisition/expropriation of areas for the insertion of the new railway line | AE | Economic costs related to the acquisition/expropriation of areas for the insertion of the new railway line |
| | Costs of investments | CI | Economic costs of the needed interventions |

(continued)

**Table 11.2** (continued)

| Clusters | Elements | | Description |
|---|---|---|---|
| Environmental aspects | Negative impact of the building site (noise + vibrations) | BS | Negative impact (noise and vibrations) due to the passage of trains |
| Transport aspects | Traffic congestion | TG | Traffic congestion due to realization/according works of the infrastructure |
| *Risks* | | | |
| Economic aspects | Lack of demand in the real estate market | LD | Lack of demand in the real estate market due to the presence of new infrastructures |
| | Possible extensions of implementation time | IT | Possible extensions of implementation time due to the conflicts arising with the local population |
| Environmental aspects | Hydro geological risk | HG | Hydro geological risk of the needed interventions |
| | Loss of biodiversity in the park | BP | Loss of biodiversity in the park due to the passage of trains in a protected area |
| Urban aspects | Dispersion settlement | DS | Risk of dispersion settlement due to the passage of trains |
| | Negative pressures on UNESCO sites | NP | Negative pressures on UNESCO sites due to the passage of trains in a protected area |

**Table 11.3** Structure of the ANP model with related maps

| *CLUSTERS* | *ELEMENTS* | *MAPS and DESCRIPTION* |
|---|---|---|
| | | ***Benefits*** |
| Economic | RE | 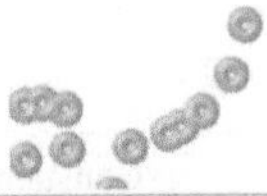 Railway Stations Buffer |
| | TL | Constant |
| Environm. | PA |  Protected Area of Magadino |
| | RA | 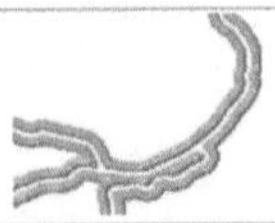 Railway Network Buffer |
| Transport | IA | Railway Stations Buffer |
| | CF | 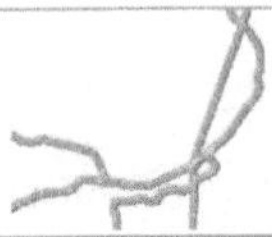 Railway Network |
| | | ***Opportunities*** |
| Economic | CE |  Railway Network |
| Transport | DT | Railway Stations |
| Urban | UC | 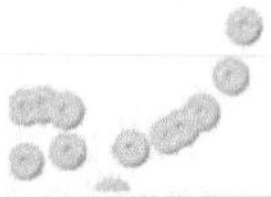 Railway Stations Buffer |
| | CT | Constant |

**Table 11.3** (continued)

| *Costs* | | | |
|---|---|---|---|
| Economic | AE | 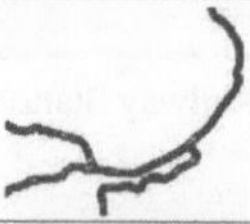 | Railway Network Buffer |
| | CI | | Constant |
| Environm. | BS | 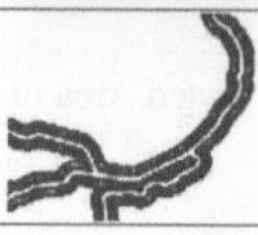 | Railway Network |
| Transport | TG |  | Road Network |

| *Risks* | | | |
|---|---|---|---|
| Economic | LD | 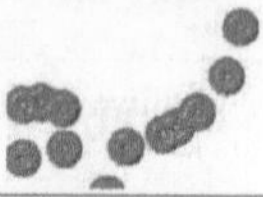 | Railway Stations Buffer |
| | IT | | Constant |
| Environm. | HG |  | Inland Waterways Network |
| | BP |  | Protected Area of Magadino |
| Urban | DS | | Protected Area of Magadino |
| | NP | | UNESCO Sites |

**Table 11.4** Pairwise comparison matrix at the cluster level for the benefits subnetwork

| Alternatives | Economic aspects | Environmental aspects | Transport aspects | Priorities |
|---|---|---|---|---|
| Economic aspects | 1 | 1/7 | 1/5 | *0.075* |
| Environmental aspects | 7 | 1 | 2 | *0.592* |
| Transport aspects | 5 | 1/2 | 1 | *0.333* |

## 11.5 Weighting and Aggregation

According to the ANP methodology described in Sect. 11.2, once the model has been structured it is necessary to develop the pairwise comparisons in order to establish the relative importance of the different elements, with respect to a certain component of the network. The comparison and evaluation phase is divided into two distinct levels: the cluster level, which is more strategic, and the node level, which is more specific and detailed. In the present application, the numerical judgments used to fill the pairwise comparison matrices were derived by a specific focus group. The focus group included different experts in the fields of transport infrastructures, environmental assessment, urban planning, economic evaluation and social sciences. The focus group had the dual purpose of helping to structure the decision problem taking into account the feedback and suggestions coming from the experts, and to compile the pairwise comparison matrices in order to come to a coherent result. Every expert was first asked to write down their individual judgments for each question. The given judgments were then illustrated and discussed in the focus group until a shared weight was achieved.

In the presented application all the calculations have been implemented using the Superdecisions software (www.superdecisions.com).

Questions such as "*Which aspects will lead to the greatest benefits associated with the transformation project? And to what extent*?" were solved by the focus group considering the cluster of the alternatives as a parent node in the Benefits subnetwork.

The judgments expressed were used to create the related pairwise comparison matrix (Table 11.4).

| | | | | | | | | | | | | | | | | | | |
|---|---|---|---|---|---|---|---|---|---|---|---|---|---|---|---|---|---|---|
| Economic aspects | 9 | 8 | 7 | 6 | 5 | 4 | 3 | 2 | 1 | 2 | 3 | 4 | 5 | 6 | **7** | 8 | 9 | Environmental aspects |
| Economic aspects | 9 | 8 | 7 | 6 | 5 | 4 | 3 | 2 | 1 | 2 | 3 | 4 | **5** | 6 | 7 | 8 | 9 | Transport aspects |
| Environmental aspects | 9 | 8 | 7 | 6 | 5 | 4 | 3 | **2** | 1 | 2 | 3 | 4 | 5 | 6 | 7 | 8 | 9 | Transport aspects |

Table 11.4 shows the pairwise comparison matrix and the main eigenvector which represents the priorities of the different aspects in the Benefit subnetwork with respect to the goal. This result highlights that environmental aspects are the most important from the Benefits point of view. According to ANP methodology, the final priority vectors that result from the comparison matrices at the cluster

**Table 11.5** Cluster matrix for the benefits subnetwork

| | Alternatives | Economic aspects | Environmental aspects | Transport aspects |
|---|---|---|---|---|
| Alternatives | 0.000 | 1.000 | 0.500 | 0.500 |
| Economic aspects | *0.075* | 0.000 | 0.500 | 0.500 |
| Environmental aspects | *0.592* | 0.000 | 0.000 | 0.000 |
| Transport aspects | *0.333* | 0.000 | 0.000 | 0.000 |

level determine the columns of the cluster matrix. Table 11.5 shows the cluster matrix for the Benefits subnetwork. The priorities of the elements that had previously been compared (Table 11.4) are shown.

Once the clusters comparison had been conducted, it was necessary to study the problem in depth through the analysis of the elements. As an example, a question submitted to the focus group was: *With reference to the evaluation of the priority of the considered projects, from the Benefits point of view, which alternative satisfies the objective "reduction in acoustic emissions" more closely? And how much more?* The judgments expressed were used to fill in the related pairwise comparison matrix (Table 11.6).

| Scenario 0 | 9 | 8 | **7** | 6 | 5 | 4 | 3 | 2 | 1 | 2 | 3 | 4 | 5 | 6 | 7 | 8 | 9 | Scenario 1 |
|---|---|---|---|---|---|---|---|---|---|---|---|---|---|---|---|---|---|---|
| Scenario 0 | 9 | 8 | 7 | 6 | 5 | 4 | **3** | 2 | 1 | 2 | 3 | 4 | 5 | 6 | 7 | 8 | 9 | Scenario 1 |
| Scenario 1 | 9 | 8 | 7 | 6 | 5 | 4 | 3 | 2 | 1 | 2 | 3 | 4 | **5** | 6 | 7 | 8 | 9 | Scenario 2 |

Once the pairwise comparison matrices had been compiled, all of the related vectors together formed the unweighted supermatrix. In this case, four supermatrices were obtained, one for each subnetwork. Table 11.7 represents the unweighted supermatrix, with reference to the Benefits subnetwork. The priorities of the elements that had previously been compared (Table 11.6) are shown.

Finally, according to the ANP methodology, the cluster matrix was applied to the initial supermatrix as a cluster weight. The result was the weighted supermatrix, which was raised to a limiting power in order to obtain the limit supermatrix, where all columns were identical and each column gave the global priority vector. In this case, four limit supermatrices were obtained, one for each subnetwork.

### 11.5.1 Final Results

Each column of the limit supermatrices obtained from the four subnetworks provides the final priority vector of all the elements being considered (Table 11.8).

The results of the complex ANP model highlight that the most important elements in the decision-making problem are: (1) reduction in acoustic emissions

**Table 11.6** Pairwise comparison matrix at the node level for the benefits subnetwork

| Reduction in acoustic emissions | Scenario 0 | Scenario 1 | Scenario 2 | Priorities |
|---|---|---|---|---|
| Scenario 0 | 1 | 7 | 3 | *0.649* |
| Scenario 1 | 1/7 | 1 | 5 | *0.072* |
| Scenario 2 | 1/3 | 1/5 | 1 | *0.367* |

**Table 11.7** Unweighted supermatrix for the benefits subnetwork

| | | Alternatives | | | Economic aspects | | Environmental aspects | | Transport aspects | |
|---|---|---|---|---|---|---|---|---|---|---|
| | | 0 | 1 | 2 | RE | TL | PA | RA | IA | CF |
| Alternatives | 0 | 0.000 | 0.000 | 0.000 | 0.105 | 0.063 | 0.051 | *0.649* | 0.122 | 0.114 |
| | 1 | 0.000 | 0.000 | 0.000 | 0.258 | 0.265 | 0.582 | *0.072* | 0.074 | 0.072 |
| | 2 | 0.000 | 0.000 | 0.000 | 0.637 | 0.672 | 0.367 | *0.279* | 0.804 | 0.814 |
| Economic aspects | RE | 0.250 | 0.500 | 0.750 | 0.000 | 0.000 | 0.000 | 1.000 | 0.875 | 0.000 |
| | TL | 0.750 | 0.500 | 0.250 | 0.000 | 0.000 | 0.000 | 0.000 | 0.125 | 0.000 |
| Environmental aspects | PA | 0.010 | 0.900 | 0.250 | 0.000 | 0.000 | 0.000 | 0.000 | 0.000 | 0.000 |
| | RA | 0.900 | 0.100 | 0.750 | 0.000 | 0.000 | 0.000 | 0.000 | 0.000 | 0.000 |
| Transport aspects | IA | 0.500 | 0.875 | 0.010 | 0.000 | 0.000 | 0.000 | 0.000 | 0.000 | 0.000 |
| | CF | 0.500 | 0.125 | 0.900 | 0.000 | 0.000 | 0.000 | 0.000 | 0.000 | 0.000 |

(environmental aspects cluster) for the Benefits subnetwork (0.163); (2) the development of the Ticino area and of its attractiveness (transports aspects cluster) for the Opportunities subnetwork (0.364); (3) the costs of the investment (economic aspects cluster) for the Costs subnetwork (0.232); and (4) the dispersion settlement (urban planning aspects cluster) for the Risks subnetwork (0.102).

### 11.5.2 Priorities of the Alternative Scenarios

The normalization of the priorities of the scenarios on the cluster of the alternatives provides the priority vector of the three considered options (Table 11.9).

Following the ANP theory, in the case of the complex network structure, it is necessary to synthesize the outcomes of the alternative priorities for each of the considered subnetworks (Table 11.9) in order to obtain an overall synthesis. Different aggregation formulas are available and the chosen formula depends on the final desired use of the results. If the purpose is to peak the best alternative, any of the different formulas will do [28]. Table 11.10 shows the final ranking of the alternative sites according to three formulas.

As is possible to notice from Table 11.10, all the available formulas converge in considering the scenario 2 as the best performing scenarios, followed by scenario 1 and finally scenario 0.

### 11.5.3 Final Decision Maps

Once the ANP questionnaire has been compiled and weights assigned to each element and cluster, the modeling system sums up all the weights and produces a three-dimensional visualization, based upon the scores given by the participants. The representation that follows is a deformation of the territory that generates a three-dimensional diagram, one for each scenario.

To obtain better comprehension on the comparison of scenarios, a slicing plane has been added to the visualization of 3D maps. This plane can be vertically moved so to work as a cursor which cuts-off the areas with lower values, i.e. those areas with less quantities of expected consequences due to the choices of actors. In this way, the slicing plane allows data to be visually selected and scenarios to be compared each other. The areas with more benefits or opportunities can be highlighted as the more preferable, while the areas with more costs and risks can be identified to better analyze the negative effects.

In the subnet of Benefits (Fig. 11.4), the system provided very different results for the three scenarios, localizing the positive effects in diverse areas. Scenario 0 shows an improvement along the rail tracks due to the reduction of noise pollution, while scenario 1 concentrates the benefits all over the Magadino protected area, which would remain outside of the zone involved in the transformation of the infrastructural system. Only scenario 2 allows a wider spread of benefits, distributing the positive effects both along the railway lines and on the park protected area, thus generating the best solution relative to this subnet. It is interesting to notice that the results of the Benefits visual map are aligned with the outcomes provided by the ANP model (Table 11.9). In fact, according to ANP priorities, the scenario 2 is the most beneficial (0.661 in the priority vector) and this finding is confirmed by the visual representation of Fig. 11.4.

Regarding the Opportunities associated with the three scenarios (Fig. 11.5), the highest peaks are in scenario 0. Those peaks are centred on the railway stations, which is the singular map associated with the "increase in connections between the Ticino and Lombardy regions". This means that the main contribution comes from the predominant importance given to the improvement of connections between the Swiss region of Ticino and the Italian one of Lombardy, identifying it as the most important element among the opinions of actors involved. The results of this evaluation strongly influences the choice of scenario 0 because it proposes a high speed connection all over the considered area. At the same time, the maps show the prevalence of scenario 0 also in the creation of new urban centralities and new job opportunities directly related to the transport improvement. It is important to underline that the results performed by the ANP, showing that the highest number

**Table 11.8** Final priorities of the elements of the model

| BOCR | Clusters | Elements | Limit priorities |
|---|---|---|---|
| Benefits | Alternatives | 0 | 0.086 |
| | | 1 | 0.110 |
| | | 2 | 0.249 |
| | Economic aspects | RE | 0.125 |
| | | TL | 0.017 |
| | Environmental aspects | PA | 0.101 |
| | | RA | 0.163 |
| | Transport aspects | IA | 0.055 |
| | | CF | 0.094 |
| Opportunities | Alternatives | 0 | 0.379 |
| | | 1 | 0.063 |
| | | 2 | 0.054 |
| | Economic aspects | CE | 0.100 |
| | Transport aspects | DT | 0.364 |
| | Urban aspects | UC | 0.028 |
| | | CT | 0.012 |
| Costs | Alternatives | 0 | 0.213 |
| | | 1 | 0.191 |
| | | 2 | 0.084 |
| | Economic aspects | AE | 0.163 |
| | | CI | 0.232 |
| | Environmental aspects | BS | 0.062 |
| | Transport aspects | TG | 0.055 |
| Risks | Alternatives | 0 | 0.324 |
| | | 1 | 0.084 |
| | | 2 | 0.066 |
| | Economic aspects | LD | 0.205 |
| | | IT | 0.099 |
| | Environmental aspects | HG | 0.023 |
| | | BP | 0.060 |
| | Urban aspects | DS | 0.102 |
| | | NP | 0.037 |

**Table 11.9** Final priorities of the alternatives under the BOCR subnetworks

| Alternatives | B | O | C | R |
|---|---|---|---|---|
| Scenario 0 | 0.194 | 0.763 | 0.438 | 0.682 |
| Scenario 1 | 0.245 | 0.128 | 0.389 | 0.178 |
| Scenario 2 | 0.661 | 0.109 | 0.173 | 0.140 |

**Table 11.10** Final ranking of the alternatives according to the different formulas

| | Additive (negative) $B-C$ | Additive (probabilistic) $B+(1-C)$ | Multiplicative $B \times (1/C)$ |
|---|---|---|---|
| Scenario 0 | −0.376 | 0.251 | 0.142 |
| Scenario 1 | −0.310 | 0.273 | 0.319 |
| Scenario 2 | 0.313 | 0.476 | 0.776 |

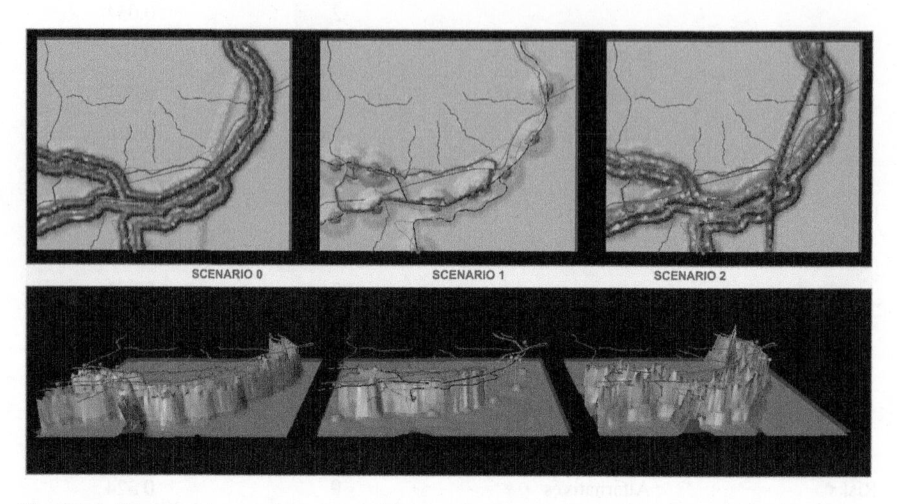


**Fig. 11.4** Visualisation of Benefits' subnet by means of 3D diagrams: a comparison among three scenarios

of opportunities is provided in scenario 0, (Table 11.9) are reflected in the maps. The analysis of Costs (Fig. 11.6) shows that scenarios 0 and 2 would be strongly characterized by costs disseminated all over the area because of the large investment required for the construction of an underground bypass. On the contrary, the costs of scenario 1 would be highly concentrated along the railway line mainly as a result of expropriation. However, the most acceptable results in terms of costs are provided by scenario 2 as emerged by the ANP method (Table 11.9).

Finally, analyzing the results related to the risk subnet (Fig. 11.7), the scenario with the lowest risks is number 1. The reason of this outcome is identified in the

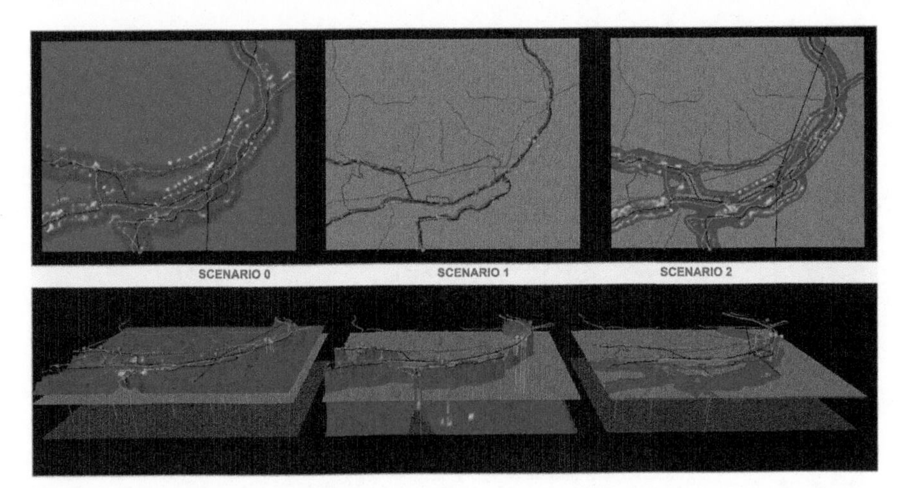


**Fig. 11.5** Visualization of opportunities' subnet

possible rebound in the housing market due to the development of the railway network. Differently, scenario 0 and 2 present many hydro-geological risks, represented by the inland waterways network. Scenario 2, in fact, is expected to be threatened by the possible extension of the implementation time due to the conflicts arising with the local population. Furthermore, a significant impact to the park area of Magadino occurs in scenario 0 due to the creation of the station in addition to the rail tunnel, while scenario 2 highlights the risk associated with general economic aspects that causes urban sprawling on the whole area in line with the results in Table 11.9.

## 11.6 Sensitivity Analysis

In order to test the model's robustness, a sensitivity analysis was performed after obtaining a ranking of the alternatives. A sensitivity analysis is concerned with the "what if" kinds of questions to see if the final answer is stable when the inputs, whether judgments or priorities, are changed. As a matter of fact, it is of special interest to see whether these changes modify the order of the alternatives.

The purpose of the sensitivity analysis is to create an explanatory process by which the Decision Makers achieve a deeper understanding of the structure of the problem. It is helpful to the analyst to learn how the various decision elements interact in order to determine the most preferred alternative and to determine which elements are important sources of disagreement among DMs and interest groups. Thus the ANP not only aids in selecting the best alternative, but also helps DMs to understand why one alternative is preferable to the other options [10].

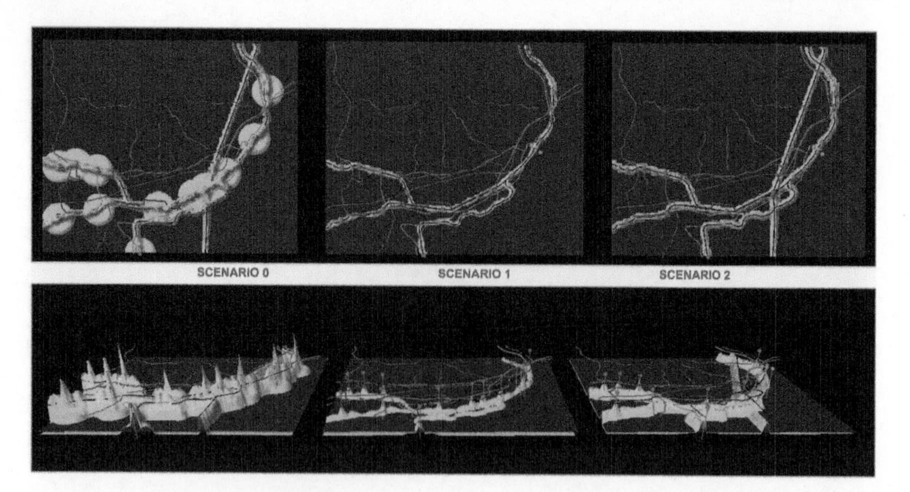


**Fig. 11.6** Visualization of costs' subnet

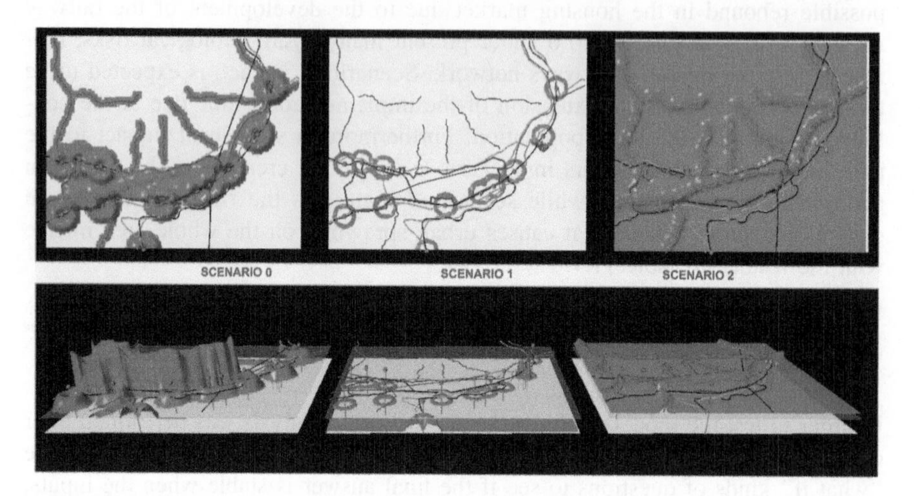


**Fig. 11.7** Visualisation of risks' subnet: comparison among scenarios

In the present chapter two different sensitivity analyses were undertaken in order to study the robustness of the model with respect to the components and interdependencies of the network. In the first analysis, the stability of the solution was studied with regard to the control criteria (BOCR) priorities. In the second, an attempt was made to verify the rank reversal of the alternatives [30] by eliminating one alternative at a time from each subnetwork of the model and thus studying the resulting final ranking, searching for potential changes.

In the first study, while measuring the sensitivity of the alternatives to the BOCR weights, an additive formulation was used, since the meaningful changes could not be obtained by a multiplicative formulation [37]. The sensitivity analysis for the four subnetworks is represented in Fig. 11.8a–d where the $x$ axis represents the changes in the weights of the control criteria and the $y$ axis represents the changes in the weights of the alternatives.

When the relationships between the Benefits dimension and the alternatives are considered it becomes clear that "scenario 2" provides more benefits compared to the other options (Fig. 11.8a). The sensitivity analysis shows that the Opportunities dimension is the most unstable subnetwork (Fig. 11.8b), since both the results and the ranking of the alternatives are very sensitive to the changes in the weight of the opportunities. The ranking of the alternatives changes from "scenario 2"–"scenario 1"–"scenario 0" (for 0 % opportunities weight) to "scenario 0"–"scenario 1"–"scenario 2" (for 100 % opportunities weight). In the Costs subnetwork (Fig. 11.8c) the alternatives are almost completely insensitive to the changes in the weight of the control criteria. Finally, the sensitivity analysis shows that the Risks dimension (Fig. 11.8d) is quite an unstable subnetwork, and one inversion in the ranking of the alternatives can be identified.

In the second sensitivity analysis, we tried to investigate the possibility of rank reversal of the alternatives [29]. This analysis eliminated one alternative at a time from the original model, and evaluated the new results. Table 11.10 thus illustrates, for each subnetwork of the model, the original ranking of the alternatives and the results arising from the elimination of the highest priority alternative. Acknowledging that rank can and should reverse under general conditions that have been recognized such as introducing copies or near copies of alternatives and criteria [33], the question is not whether rank should be preserved [39], but whether or not the assumption of independence applies [33]. As it is possible to see from Table 11.11, the rank is preserved, with a small exception for the risks subnetwork where the two alternatives rank very similarly; it is thus possible to conclude that the final result of the model is stable.

## 11.7 Conclusions

This chapter illustrates the application of a complex ANP method to support the decision process regarding the different projects of the railway in the Bellinzona area. At the same time, this article presents a new approach to the integration of the modelling system for the spatial visualization of the ANP.

The ANP methodology is able to take into consideration both tangible and intangible criteria and considers the relationships between these in a systematic manner. This is particularly important for assessing the processes of urban and territorial transformation, as the case presented here.

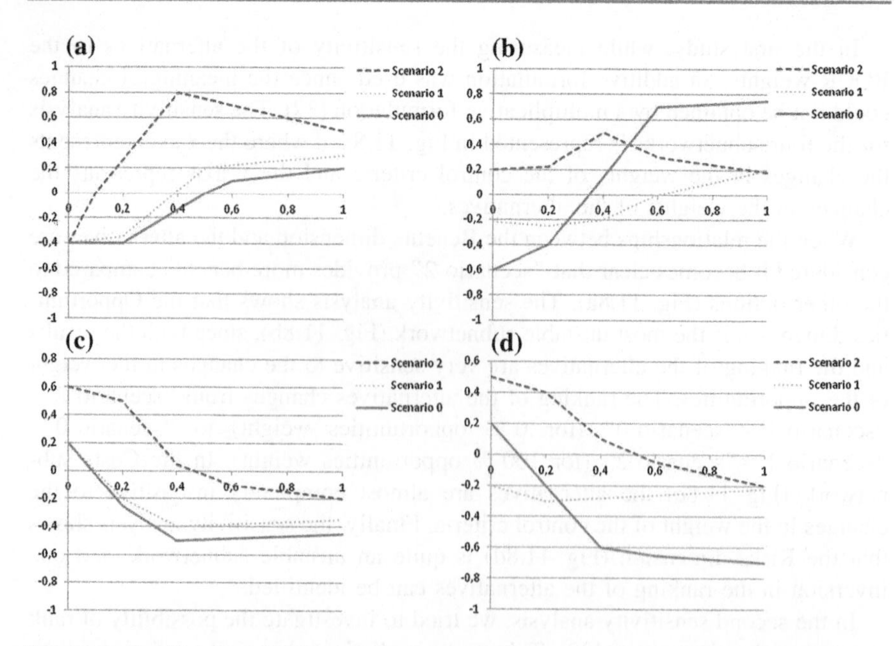


**Fig. 11.8** Sensitivity analysis for each subnetwork using the additive (negative) formula

The ANP allows the most important elements of the decision to be highlighted through a transparent and traceable process, thus facilitating deliberation.

A weakness of this methodology could refer to the assignment of weights to clusters and nodes. In fact, this procedure could create misunderstandings due to a lack of ability of non-expert users to understand their meaning. The use of ANP to study alternative planning solutions in real decision arenas helped identify areas for potential improvement [11] in this sense: a more visual grammar can generate a common basis for sharing information and allowing discussions. Therefore, the indexes, weights and rankings from ANP must be an object of discussion as well as their results. Furthermore, the large quantity of data to manage during the decision process has highlighted the necessity to filter items in order to better identify and isolate core features.

The results of the analysis performed show that the ANP-BOCR model is suitable to represent a real world problem. The technique provides the means to perform complex trade-offs on multiple evaluation criteria, while taking the DMs preferences into account. The main drawback in the practical application of the ANP is a consequence of the complexity of the decision issue being analysed. For example, the ANP prescribes a large number of comparisons that occasionally become too complex for DMs to understand if they are not familiar with the method. Hence, a great deal of attention should be devoted to the wording of the questionnaires and the comparison process should be helped by a facilitator (Aragonés-Beltrán et al. 2010). The evaluation process has been supported by the use of interactive visualisations, which were effective in the creation of a common

**Table 11.11** Sensitivity analysis with respect to the rank reversal of the alternatives

| Subnetwork | Priority of the alternatives | Original ranking | Eliminated alternative | New priorities | New ranking |
|---|---|---|---|---|---|
| Benefits | 0: 0.194<br>1: 0.245<br>2. 0.661 | 2 > 1 > 0 | 2 | 0: 0.347<br>1: 0.653 | 1 > 0 |
| Opportunities | 0: 0.763<br>1: 0.128<br>2. 0.109 | 0 > 1 > 2 | 0 | 1: 0.573<br>2: 0.427 | 1 > 2 |
| Costs | 0: 0.438<br>1: 0.389<br>2. 0.173 | 0 > 1 > 2 | 0 | 1: 0.660<br>2: 0.339 | 1 > 2 |
| Risks | 0: 0.682<br>1: 0.178<br>2. 0.140 | 0 > 1 > 2 | 0 | 1: 0.494<br>2: 0.506 | 2 ≈ 1 |
| BOCR | 0: 0.251<br>1: 0.273<br>2. 0.476 | 2 > 1 > 0 | 2 | 0: 0.444<br>1: 0.556 | 1 > 0 |

mental model among the actors involved. Furthermore, the use of visual communication provided the basis for the sharing of information and the creation of individual awareness and enhanced the discussion between the parties.

However, there are still a number of opportunities for expanding the study and for validating the results obtained. First, it would be of scientific interest to weight the BOCR categories by implementing the evaluation model by means of the strategic criteria [31]. Second, the model could be combined with a Costs-Benefits Analysis in order to develop an overall assessment of the transformation project impacts [36].

In conclusion, the methodology adopted was successful in structuring the complex planning context, communicating the stakeholders' perspectives, improving the stakeholders' commitment and their perception of being involved, enhancing transparency in the decision-making process and thus increasing the acceptance of the proposed solutions.

## References

1. Abastante F, Bottero M, Lami IM (2012) Using the analytic network process for addressing a transport decision problem. Int J Anal Hierarchy Process 4:41–60

2. Abastante F, Lami IM (2012) A complex analytic network process (ANP) network for analyzing Corridor24 alternative development strategies. Paper presented at the 2nd international conference on communications, computing and control applications (CCCA' 2012), Marseilles, France, 6–8 Dec 2012
3. Agarwal A, Shankar R, Tiwari MK (2006) Modelling the metrics of lean, agile and leagile supply chain: an ANP-based approach. Eur J Oper Res 173:211–225
4. Andrienko G, Andrienko N, Jankowski P, Keim D, Kraak MJ, MacEachren AM et al (2007) Geovisual analytics for spatial decision support: setting the research agenda. Int J Geogr Inf Sci 21(8):839–857
5. Bottero M, Lami IM, Lombardi P (2008) Analytic network process: la valutazione di scenari di trasformazione urbana e territoriale. Alinea, Firenze
6. Bottero M, Lami IM (2010) Analytic network process and sustainable mobility: an application for the assessment of different scenarios. J Urbanism 3:275–293
7. Bottero M, Mondini G (2008) An appraisal of analytic network process and its role in sustainability assessment in Northern Italy. Int J Manage Environ Qual 19:642–660
8. ETH (2011) Draft Dossier Ticino–Lombardia. Zurich July 2011, Mimeo
9. Figueira J, Greco S, Ehrgott M (eds) (2005) Multiple criteria decision analysis: state of the art surveys. Springer, New York
10. Khan S, Faisal MN (2008) An analytic network process model for municipal solid waste disposal options. Waste Manag 28:1500–1508
11. Lami IM, Masala E, Pensa S (2011) Analytic network process (ANP) and visualization of spatial data: the use of dynamic maps in territorial transformation processes. Int J Anal Hierarchy Process (IJAHP) 3:92–106
12. Marina O, Masala E, Pensa S, Stavric M (2012) Interactive model of urban development in residential areas in Skopje. In: Leduc T, Moreau G, Billen R (eds) Usage, usability and utility of 3D city models. EDP Sciences, Les Ulis
13. Masala E (2012a) ETH 13 Dec 2011—Computational assessment workshop. https://www.youtube.com/watch?feature=player_detailpageandv=PSnu_Ti66VU. Accessed 8 June 2012
14. MacEachren AM, Taylor DR (1994) Visualization in modern cartography. Pergamon Press, Oxford
15. MacEachren AM, Gahegan M, Pike W, Brewer I, Cai G, Lengerich E et al (2004) Geovisualization for knowledge construction and decision-support. Comput Graphics Appl 24(1):13–17
16. Malczewski J (2006) GIS-based multicriteria decision analysis: a survey of the literature. Int J Geogr Inf Sci 20(7):703–726
17. Niemura MP, Saaty TL (2004) An analytic network process model for financial-crisis forecasting. Int J Forecast 20:573–587
18. Neaupane KM, Piantanakulchai M (2006) Analytic network process model for landslide hazard zonation. Eng Geol 85:281–294
19. Pensa S (2012) InViTo partecipatory process test, Torino. https://www.youtube.com/watch?v=EVpm1EW7z-sandlist=PLA68D9CE96846CD66andindex=1. Accessed 22 Feb 2013
20. Pensa S (2013) InViTo, geoVisualizzazione Interattiva a Supporto dei Processi di Decisione Territoriale. Ph.D. candidate thesis, Politecnico di Torino
21. Pensa S, Masala E, Lami IM (2013) Supporting planning processes by the use of dynamic visualization. In: Geertman S, Toppen F, Stillwell J (eds) Planning support systems for sustainable urban development, Springer, Heidelberg, pp. 451–467. ISBN: 9783642375323
22. Pensa S, Masala E, Marietta C (2011) The effects of decision-making on urban form: a tool for supporting planning processes. In: Pinto NN, Tenedorio JA, Santos M, Deus R (eds) Proceedings of the 7th international conference on virtual cities and territories, Lisbon
23. Piantanakulchai M (2005) Analytic network process model for highway corridor planning. Paper presented at the international symposium on the analytic hierarchy process (ISAHP 2005), Honolulu, 8–10 July 2005

24. Promentilla MAB, Furuichi T, Ishii K, Tanikawa N (2006) Evaluation of remedial countermeasures using the analytic network process. Waste Manag 26:1410–1421
25. Roy B, Bouyssou D (1993) Aide multicritère à la décision: Méthodes et cas. Economica, Paris
26. Saaty TL (1980) The analytic hierarchy process, planning, priority setting, resource allocation. McGraw-Hill, New York
27. Saaty TL (2001) The analytic network process. RWS Publications, Pittsburgh
28. Saaty RW (2003) Decision-making with the AHP: why is the principal eigenvector necessary. Eur J Oper Res 145:85–91
29. Saaty TL (2005) Theory and applications of the analytic network process. RWS Publications, Pittsburgh
30. Saaty TL (2006) Rank from comparisons and from ratings in the analytic hierarchy/network processes. Eur J Oper Res 168:557–570
31. Saaty TL, Vargas LG (2006) Decision making with the analytic network process. Springer Science, New York
32. Saaty TL, Ozdemir MS (2008) The encyclicon: a dictionary of applications of decision making with dependence and feedback based on the analytic network process. RWS Publications, Pittsburgh
33. Saaty TL, Sagir M (2009) Extending the measurement of tangibles to intangibles. Int J Inf Technol Decis Making 8:7–27
34. SiTI (2012) InViTo: interactive visualization tool. http://www.siti.polito.it/index.php?id=1andt=tpl_6andidp=191. Accessed 10 Dec 2012
35. Treu MC, Russo G (2009) La via delle merci. Il Sole 24 ore libri, Milano
36. Tsamboulas D, Mikroudis G (2000) EFFECT-evaluation framework of environmental impacts and costs of transport initiative. Transportation Research Part D: transport and environment 5(4):283–303
37. Tuzkaya G, Onut S, Tuzkaya UR, Gulsun B (2007) An analytic network process approach for locating undesirable facilities: an example from Istanbul, Turkey. J Environ Manage 88:970–983
38. Tuzkaya G, Onut S (2008) A fuzzy analytic network process based approach to transportation-mode selection between Turkey and Germany: a case study. Inf Sci 178(15):3133–3146
39. Tversky A, Slovic P, Kahneman D (1990) The causes of preference reversal. Am Econ Rev 80(1):204–215
40. Ulutas BH (2005) Determination of the appropriate energy policy for Turkey. Energy 30:1146–1161
41. Van den Brink A, Van Lammeren R, Van de Velde R, Dane S (2007) Geo-visualization for participatory spatial planning in Europe: imaging the future, vol 3., Mansholt Publication SeriesWageningen Academic Publishers, Wageningen

## Reference Sites

1. www.superdecisions.com

# 12 Analytic Network Process, Interactive Maps and Strategic Assessment: The Evaluation of Corridor24 Alternative Development Strategies

Francesca Abastante, Felix Günther, Isabella M. Lami, Elena Masala, Stefano Pensa and Ilaria Tosoni

**Abstract**

The development of strategic transport infrastructure needs to be accompanied through dedicated processes, enabling both the discussion and sharing of the expected positive and negative impacts and externalities. According to the different scales of intervention, spatial strategies can assume different shapes and require the involvement of a different range of experts and stakeholders. The topic of involvement and long-lasting collaboration, in these cases, touches several dimensions: the actual physical and geographical distance between

F. Abastante (✉) · I. M. Lami (✉)
Department of Regional and Urban Studies and Planning (DIST),
Politecnico di Torino, Corso Massimo D'Azeglio 42, 10123 Turin, Italy
e-mail: francesca.abastante@polito.it

I. M. Lami
e-mail: isabella.lami@polito.it

F. Günther · I. Tosoni
Institute for Spatial and Landscape Development, Chair for Spatial Development,
ETH Zürich, Wolfgang-Pauli-Str. 15 (HIL H 37.4), 8093 Zürich, Switzerland
e-mail: fguenther@ethz.ch

I. Tosoni
e-mail: ilaria.tosoni@gmail.it

E. Masala · S. Pensa
SiTI—Higher Institute on Territorial Systems for Innovation,
via Pier Carlo Boggio 61, 10138 Turin, Italy
e-mail: elena.masala@polito.it

S. Pensa
e-mail: stefano.pensa@polito.it

I. M. Lami (ed.), *Analytical Decision-Making Methods for Evaluating Sustainable Transport in European Corridors*, Sxi 11, DOI: 10.1007/978-3-319-04786-7_12,

some of the actors; different cultural backgrounds and languages; the difficulty of reading outputs. The study presents a Collaborative Assessment Workshop designed in order to allow different forms of interaction between the participating actors and enabling them to contributing referring to both their knowledge as local stakeholders and experts' skills. The first step of the assessment is structured through an approach integrating Analytic Network Process (ANP) and Interactive Visualization Tool (InViTo) in order to create a shared basis for generating discussion. The second step is constituted by a Collaborative Assessment assuming the perspective of strategy design processes as peer learning practice. The Collaborative Assessment Workshop aims at identifying a shared spatial and infrastructural development strategy for the regions connected through the railway Corridor24 Genoa–Rotterdam.

## 12.1 Introduction

Spatial development strategies are complex design objects requiring, in order to be effective, the active engagement of several actors [19–21]. Such engagement concerns a variety of resources among which time and knowledge appear to be the most relevant. According to the different scales of intervention, spatial strategies can assume different shapes and require the involvement of a different range of experts and stakeholders [5]. In addition to this, in the case of large-scale spatial strategies, the amount of information to be managed and the number of actors to be involved is such that approaching this task requires a dedicated organization to accompany the strategy's design process. Not many examples of such enterprises are available to the analysis of scholars and practitioners [5, 20] and therefore this is a challenging field of research for new forms and instruments of intervention.

All these issues present themselves in the case of the CODE24 project. The intervention's perimeter is defined by the European rail corridor Rotterdam–Genoa, central north-south infrastructure of this region, and all territorial entities (ports, regions, cities, business companies, communities), which relate to this infrastructure for their functioning. The area of interest therefore has a longitudinal extension of more than 1,000 km, covers a surface of more than 30,000 km$^2$ and concerns a population of more than 70 million inhabitants. In this respect the area presents great questions concerning both the feasibility and the possible form of a common strategy (contents, detail, outlook) and the kind of decision-making process necessary to reach this goal. It is in fact acknowledged that such attempts cannot be considered simple planning episodes, but, in order to produce concrete outcomes, strategy design needs to be a continuous process, which needs to be established and maintained through time by a designing community [20].

The topic of involvement and long-lasting collaboration in these cases, touches several dimensions: the actual physical and geographical distance between some of the actors, which could loosen the need for maintaining operative collaboration due to different interests, contingent situations and priorities; different cultural backgrounds and languages which also played a crucial role in setting the partnership's framework and developing instruments to bridging the gaps and difficulties.

For these reasons the Collaborative Assessment Workshop has been designed by the ETH Zurich's and SiTI's researchers in order to allow different forms of interaction between the participating actors and enabling them to contribute referring to both their knowledge as local stakeholders and experts' skills. The assessment foresees, in fact, four different stages of interaction with the content materials and the other participants:

- Remote individual work.
- Structured sessions steered by a moderator and facilitators.
- Semi-structured sessions with the help of a facilitator.
- Unstructured and open discussion.

During the different sessions the participants are asked not only to evaluate and provide feedback to materials presented by the organizers, but also encouraged to present their own contributions and ideas.

The Collaborative Assessment Workshop regarding the whole Rotterdam–Genoa corridor has been conducted in two sessions of one and half days, but it came at the end of a larger preparatory process, composed of meetings, focus groups and workshops. The preparation has been important to generate the right collaboration climate for the implementation of the procedure and to allow the participants to be familiar with the issues and the material, which would then have been object of the assessment.

The workshop was expected to stimulate the discussion and the sharing of relevant ideas about the future of the corridor. The aim was to support the stakeholders' acknowledgment of the strategic spatial components of a common strategy and also to pursue the local, regional and macro-regional spatial impacts connected to the alternative strategic perspectives. In this respect a great field of research and exploration has been that of visual tools and codes supporting the collaborative design process, both in terms of possibility of direct interaction with the project dimension, development of a common symbolic language and forms of visual communication towards the general public.

Ultimately, the future management, implementation and update of the strategy as reinforcement and empowerment of the established collaboration, which are connected to these issues, were therefore also discussion topics.

## 12.2 Transport Projects, Territorial Impacts and Decision-Making Processes

Major infrastructure are territorial works for two reasons: (1) because they structure the territory at their own level (national or international); (2) because the relative paths and access points appear in turn as opportunities and risks to the regional and local levels of government both in the places crossed by them or included in the fields of externalities which they give rise to, or change [14].

In this sense, it seems appropriate to highlight some peculiarities of the transport infrastructure. They are characterized by three types of indivisibilities [12]:

(a) Indivisibility of scale (technical). A road, a railway or an airport must have the technical dimension and the equipment needed to meet the purposes for which they are made. In general, the transition to a higher level of infrastructure leads to larger jumps in continuity.
(b) Temporal and financial indivisibility. An infrastructure project can be used only if it is reasonably complete or, at least, none of the its component parts can be used if these parts are not complete, and if they do not have a minimum of significance in technical and economic terms.
(c) Indivisibility of system. No infrastructural work can produce effects and benefits if it is not somehow embedded in a system of infrastructure, both of similar or different type, so as to perform its essential function as an interconnecting element.

With respect to (a) and (b), one could argue that, in fact, a part of a transport infrastructure is often made available to traffic before the entire complex is completed. However it should be noted that [28]:

- The parts that have been built, if they are to be used, must have a minimum of technical consistency and a clear economic significance.
- It is unlikely that the effects produced by a single, limited fraction of the whole work represent a proportional sub-multiple of the overall benefits in view of which the work itself has been designed and is being built.

Compared to (c), it has to be mentioned that much of the literature of urban planning and geography has shown that increasing the complexity of interconnection levels within infrastructure networks causes urban and territorial impacts to become greater.

Interconnection can play a primary role, independent and highly unique, in the processes of structuring settlements, producing new forms of urban patterns, that simple infrastructure connections (connection between two nodes) would not be able to generate. The latter, in fact, seem quite likely to strengthen or extend the existing settlement structures in incremental forms, only partially innovative. Therefore the fundamental difference between the two morphogenesis is generated by their different level of complexity. The first one, centred on interconnection, is greater and synergistic, based on positive sum games (creation of externalities during the process). The second one, simple connection, produces a very limited morphogenetic process, based on zero-sum games (exploitation of given externalities) and is generally opportunistic towards the home network and the organization of the territory [14].

From the economic-financial perspective, transport infrastructures are characterized by a deferred profit over time and which is often hardly noticeable in the form of revenue allowing the recovery of the invested capital and financial interests. In this sense four types of profitability can be identified [27]:

- Transport infrastructure produces an overall and social profitability: the opening of new opportunities for trade, movement of people, production, residential and tourism locations and new opportunities for specialization and territorial division of labour. A wide category of effects is likely to occur in time in the area directly affected by the infrastructure, ranging between changes in income, employment and productivity, as well as in income regarding the overall interregional and national context which is a direct part of the affected area.
- A second type of profitability is linked to the effects on land use within the most immediate influence of the infrastructure. The location of industrial sites and residential forms a belt around infrastructure or its stations and junctions. The interest in these areas is mainly a function of the distance from the infrastructure and the facilitating accessibility to the latter. This land value pattern proves to have no or little importance in the context of the economic effects of infrastructure development.
- A third type of profitability is represented by the "multiplier" effects on income and employment arising from the investment.
- The fourth category of effects, or profitability, is the one that directly involves transportation costs, and the impact of the infrastructure in the determination (and reduction) of these costs.

Finally, it has to be mentioned that a large transportation infrastructure is in itself highly selective, it improves traffic conditions in some areas and worsen them in others. A large transport infrastructure therefore opens new systems of inequalities and conflicts within the social and territorial fabric, which, due to spatial dispersion, used to perceive itself as a homogeneous entity [51].

For all these reasons the development of strategic transport infrastructure needs to be accompanied through dedicated processes, enabling both the discussion and sharing of the expected positive and negative impacts and externalities. The participation of the stakeholders at all levels in the decision-making allows to explore all issues at stake, and develop accompanying measures and projects enabling a better embedment of the new infrastructure in the existing context. The aforementioned Collaborative Assessment Workshop has been designed in order to provide a suited setting to this crucial part of the decision-making.

## 12.3 Methodological Framework

### 12.3.1 MCDA and Visualization

The illustrated application can refer to the family of the Problem Structuring Method approaches (PSMs). In fact, the PSMs are usually applied in complex situations characterized by multiple actors and perspectives, incommensurable and/or conflicting interests, important intangibles and key uncertainties [41].

It has been frequently suggested that the construction of complex problems is a result of the process of problem structuring rather than being given a starting point. It may therefore be better to talk about different aspects or dimensions of a problem situation, rather than different types of problems [32]. In this sense, the PSMs enable participants to clarify their predicaments, converge on a potentially mutual problem on which action could be taken or issue within it and agree on commitments that will at least partially be resolved. To be effective a PSM should enable several alternative perspectives to be brought into conjunction, be cognitively accessible to actors with different background, operate iteratively so that the problem representation adjusts to reflect the stage and state of the discussion and to permit partial or local improvements [33].

The literature reports a variety of problem structuring methods [29, 33, 38] as for example: Strategic Options Development and Analysis (SODA), Soft Systems Methodology (SSM), Strategic Choice Approach (SCA) Robustness Analysis, Drama Theory, Interactive Planning [2], Scenario Planning [46], SWOT analysis [58] etc.

The cited approaches have not simply the possibility of addressing awkward problems in common but the chance to integrate visual representation (both high-tech and low-tech) in group model-building [57].

In the presented ANP/InViTo approach, visual representation plays a key role in the content and process of collaboration [31], helping people involved to "get on the same page" [57] and to gain a collective insight [6] about the issues involved.

The visual representation used in collaborative problem solving and group model-building plays a significant role in facilitating and shaping consensus when they function as *boundary objects* [8]. A boundary object is defined as "an object adaptable enough to be interpreted differently by people with differing expertise without losing a coherent identity spanning across the words" [53].

Visual products of facilitated processes can serve as *boundary objects* when they are (McKenzei and Winkelen 2001):

1. Tangible two-dimensional or three-dimensional shared representations.
2. They portray salient dependences and relationships among participants' objectives, expertise, decision and actions.
3. They can be modified by input from every participant. In this sense if an object is transformable, then anyone involved has the possibility to manipulate and alter the representation to show the consequences of the perceived dependences more clearly.

In order to facilitate the decision process, it is not sufficient to apply a good visualization tool. It is necessary to have "good decision bones", to structure the decision problem in a simple and effective way to capture the complexity of the reality. Mention has to be made to the fact that the view of strategic choice is essentially about choosing in a strategic way rather than at a strategic level. The concept of strategic choice here illustrated is about the connectedness of one decision with another rather than about the level of importance attached to one decision in relation to others [18].

To achieve this objective, MCDA can be pivotal in complex decision processes. As the MCDA methods are countless, it is useful and necessary to deeply reflect on the most suitable method that can be adapted for the decision context [42]. In order to structure the decision process regarding the evaluation of alternative development strategies for the Corridor24, we chose to apply the Analytic Network Process (ANP) methodology for several reasons.

First, the type of results the ANP methodology assigned are numerical values to each potential action. Moreover the ANP methodology is able to produce a list of *k-best* actions to be further analysed by the people involved. Second, the original performance scale of the ANP method, the Saaty's fundamental scale of absolute numbers [44] has all the properties required for a correct application. Hence there is no need to transform or codify the original scale, which could cause the rise of arbitrary transformation that could, in turn, affect the process as a whole. Third, the software tool involved (www.superdecisions.com) and the interaction protocol are compatible with the way people think and the usefulness of the results obtained. Finally, the ANP is a simple and understandable methodology even by those who are not experts in the decision process and it is suitable to be applied jointly with visual representations in real time during the workshops.

Therefore, the following research focuses on the use of the Analytic Network Process (ANP) methodology combined with the Interactive Visualization Tool (InViTo) for the visual assessment of spatial issues. InViTo is a visual method to communicate spatial information, which aims at improving the understanding of spatial data in decision-making processes through the exploration of alternative strategy options. [24, 37–39]. The application of InViTo is expected to provide a qualitative methodology to communicate numerical data. The application of ANP asks users to evaluate different spatial and non-spatial aspects by means of Saaty's fundamental scale. Therefore, ANP process gives back a quantitative assessment of problems, which can now be easily compared. However, the comparison of matrices is not so immediate, especially when they include more than four or five items at once (see Chap. 10, Masala and Pensa). For this reason, InViTo proposes a method to translate a large amount of numbers into visual representations, which can act as a synthesis of all the considered aspects.

The visual representations presented in the paper act as *boundary objects*. They are a tangible representation of dependences across disciplinary, organisational and cultural lines that all participants can modify. In this perspective InViTo allows transforming a shared representation, which help the participants to navigate in less threatening and risky ways to a shared understanding of what they could do individually or collectively do [8].

### 12.3.2 Collaborative Assessment

As stated in Chap. 9 (Tosoni and Guenther), the approach behind the Collaborative Assessment procedure has been consolidating over time assuming the perspective that sees collaborative process of constructing spatial strategies as peer learning processes [11, 22, 52]. Those processes are designed as occasions to share knowledge and skills possessed by the different stakeholders, which are therefore directly involved in the strategy's design process. Thus, newly shared knowledge together with concrete intervention proposals are the main outcomes of such activities.

The Collaborative Assessment Workshop has been designed as the core step of a wider assessment process. The overall process aims at exploring the territories (included in the action's perimeter) through the eyes of the most relevant actors, consolidating a partnership and, at the same time, defining a common working path for the strategy's implementation. The process has been developed following the suggestions of the so-called Action Planning approach [47], which states that the solution of complex problems, such as spatial development strategies, require tailor-made organizations to provide the needed environment and scheduling for the knowledge sharing to take place. The approach also considers the evaluation of project alternatives as the key element to a better understanding of the problems at stake [48]. Problems and solutions are, in fact, incrementally honed through several cycles of assessment [30].

The process necessarily takes place in a social context, characterized by very different kinds of actors and agents, which bear different competences, skills, beliefs and interests [13, 15, 16, 17, 26]. In this perspective the decision-making process is a dialogue and interactive practice [21, 36, 49] which, in order to grant the needed consensus and social support, needs to be inclusive and open to the different contributions [50, 55].

The dialogue element can, nevertheless, cause difficulties with regard to confrontation and conflicts, which depends on both content matters; different opinions about alternative project solutions, but also confrontation between different disciplinary approaches or role conflicts between actors competing for a leadership position in the process [7, 54]. As a result planners involved in such processes need to act assuming different roles at the same time [17, 23] mediators in the confrontation and negotiation phases; facilitators during the discussions and designers in the accompanying the phases of transposition of the discussion's outcomes into project components.

The Collaborative Assessment has been developed as an instrument supporting these features; in particular it assumes the perspective of strategy design processes as peer learning practice [11, 22] where the active involvement of the stakeholders is essential to generate crucial knowledge for designing the joint actions. Starting from these considerations the assessment process foresees different activities aimed at initiating and consolidating the collaboration between the participants. A first step includes critical mapping of the main issues as a first acknowledgement of the several views and ideas about problems and their solutions. This activity starts at the local level and then embraces the area of intervention as a whole. Design intervenes as an exploration tool at first during the mapping phase (through the assembling of common overviews, [47]) and, once the conditions are created, as field of development of comprehensive project alternatives. The joint evaluation of the project alternatives is then followed by the discussion of a common work plan.

The devised path aims at encouraging, in particular, two key aspects of the participation of the actors:

- The introduction and enhancement of their knowledge about the territory in the process of issues' emergence.
- The facilitation of their contribution, not only to the evaluation of design alternatives, but also to develop their own proposals.

In order to achieve these goals, the participants have to be familiar with the contents of the assessment, which means the key elements have to be discussed prior to the assessment workshop. A common language is needed both semantically and visually. It has also to be said that in cases such as those of large scale interventions, the amount of information to be processed is significant and it is important to provide the actors with the all the needed sources, but at a same time a smart way of simplifying the dense layers of content has to be found in order to make the issues approachable. It is also important to provide a wide range of interaction settings either accompanying the design moments, or supporting the evaluation task. Also to achieve this aim the Collaborative Assessment workshop has been designed to include different modules: structured assessment session (through ANP or remote or live questionnaires), free discussion sessions, working in small groups, individual contributions.

Concerning the content matter of the assessment intervention, different ways of approaching spatial planning and design were provided: draft projects to be discussed in small groups, input packages reporting the outcomes of the several assessment meetings and anticipating trial strategies to be discussed, comparison of test plans, composition of desired outcomes choosing among a set of given project components. All these trials allowed to identify the appropriate proceedings in order to foster a competent, open and creative knowledge sharing and formation. The final setting is based on a simple game-like strategy design where information is simplified in small packages to be manageable for the actors. The simple comparison of the project's comprehensive alternatives proved to lead to a too radical simplification of the issues at stake. In order to avoid this problem, strategy cards have been designed presenting development scenarios though visual representations, lists of projects, budget information and expected outcomes. The game setting was chosen to create the right climate stimulating both a positive sharing attitude among the participants and a relaxed competition between ideas.

Through this scheme, repeated in several cycles, it has been possible to undertake a reasoned and critical mapping of the various components that inform the spatial strategy and to reach a shared definition of the guiding principles of the strategy itself.

## 12.4 Application

Based on a problem-oriented and not on geographical approach, Corridor24 has been divided into 9 subregions, which present analogue problem backgrounds:

1. Betuwe Region.
2. Koln area.
3. MiddleRhein.
4. Frankheim region.
5. UpperRhein.
6. NorthWest Switzerland.
7. Central Switzerland.
8. Ticino–Lombardia.
9. Ligurian range.

For each region alternative strategies have been explored and drafted. The key components of the strategies reflect the clusters and nodes of the ANP model as well. The participants have been asked to rate their preferences concerning the key elements, which should shape the common strategy. The ANP structured discussion provided a set of preferences regarding thematic issues and possible project components, which have been formulated as principles guiding the design of the common strategy. Starting from the results of the ANP session, the participants have been then asked to verify the expressed preferences in view of their spatial representation in terms of projects and spatial development perspectives. In the Strategic Assessment session they have been divided in 2 groups (of max 6/7 experts) and asked to build-up their proposal by selecting between strategy—packages produced for each region of the corridor. Two teams had to choose between a minimum of 2 and a maximum of 4 alternatives for each region. These alternatives for the single macro-regions were pictured on a card including: a list of key projects, including costs and total budget; a brief description of the key issues that the strategy aims to tackle; a brief description of the strategic actions foreseen by the strategy; a visualization of the ANP elements included in the strategy. With the support of a facilitator, each group had to compose its preferred strategy, also by creating their own additional proposal.

On the basis of the final plenary discussion, the participants agreed on a first draft of strategic recommendations addressed to the project partnership, of national and local political representatives.

### 12.4.1 Process of the Evaluation

Although the workshop itself lasted one day and a half, the preparation process started 8 months earlier (Table 12.1). In fact, the evaluation models (and the evaluation method used to structure the decision process) are subject to a validation process that involves four steps: conceptual, logical, experimental and operational validation [25]. The aim of the validation process is to verify if the key issues have been appropriately considered [56]. The first part of the evaluation process took place in the experimental validation step. This was divided in turn into two pilot tests: a test with real actors, experts and researchers and, according to the feedbacks received, a second internal test with experts in decision processes, planners and visualisers in order to test possible changes. The experimental validation consists in testing the model, using experimental data and examples, in order to show if it is able to provide the expected results, before applying it in a real decision context [35].

The application described in this chapter was the last one of the whole project, so it gained the maximum benefits from the previous experiences of five workshops held between 2011 and 2013 (see Chap. 6, Lami).

The pilot test with real actors, experts and researchers, was very useful to improve the structure of the workshop itself from different points of view: the choice of the type of ANP network the explanation of the questions, the type of visualisations and the interaction with the collaborative assessment. Previous to each workshop with DM, an online questionnaire was circulated.

The online submission of the questions generated through the ANP model represents an unconventional use of this methodology, but it has been chosen for the following reasons: (1) it allows the start of the plenary discussion during the workshop and interfacing with informed actors; (2) it allows the reduction of problems arising from the process of social influence. In fact every actor has the ability to answer the questions in his sphere of autonomy without being influenced by other participants, who may have a greater charisma and, otherwise, monopolize the votes. This aspect could be seen as a weakness however; while it is true that each person is free to respond without any influence, it is also true that the actor may not be sufficiently informed about the facts required and could therefore only answer the question incompletely. Furthermore, as the time available for workshops was limited, the online questionnaire helped by collecting information that would otherwise have required a whole workshop session, which was not possible. On the other hand, the answers given to the online questionnaire provide a useful control parameter for testing the influence of the live discussion of the participants' points of view: as it is possible to distinguish whether the 'change of mind' depends on a better understanding of the issues and questions or on the interaction with the other actors.

An example of the preliminary decision areas defined by the information collected through the online questionnaire is provided in Fig. 12.1. In the reported question, participants were asked to give a weight (using the Saaty's scale) to the

**Table 12.1** The mains steps of the evaluation process

| Time | Activity | Actors (n°) | ANP network | Visualisations | Collaborative assessment |
|---|---|---|---|---|---|
| July–Aug. 2012 | Definition of three alternative strategies for the whole Corridor and construction of the complex strategic ANP network | | Two levels ANP. Top level: single strategic network including BC nodes and strategic criteria. Second level: BC network | 3D symbolic visualisations | Selection of the strategies's components and features; Definition of the reference scenarions |
| Sept. 2012 | Pilot test on the case study with real actors and experts (at Genoa Port Authority) | 9 | Complex strategic ANP network | 3D symbolic visualisations | Definition of the tasks for the experts groups work |
| Oct.–Dec. 2012 | Definition of a new ANP network, new visualizations, new application of the collaborative assessment | | Simple ANP network to rank the criteria | 2D histograms describing the distribution along the corridor of the effects of DMs'choices | Review of the strategic assessment setting: from comparison of strategies to comparison of projects |
| Jan. 2013 | Pilot test on the case study with a small number of experts (at ETH) | 10 | Simple ANP network to rank the criteria | 2D radial charts distributed along the corridor | Test with game-like and role playing setting |
| Jan. 2013 | Online questionnaire (in German and English) | 10 | | | Work on the language of the assessment: wording, meanings, and references |
| Feb. 2013 | Aggregation of the weights and identification of the most controversial questions | | The elicitation of the weight was done by the "Majority method" | Bubble charts | Real-time collection of inputs for the group discussion |
| 6–7 Feb. 2013 | Workshop with real DMs, internal and experts (at ETH ValueLab) | 10 | Simple ANP network to rank the criteria | One interactive 2D radial charts of BC nodes | Facilitation of the interactions in the groups using the strategy cards, maps and guidelines for questions |

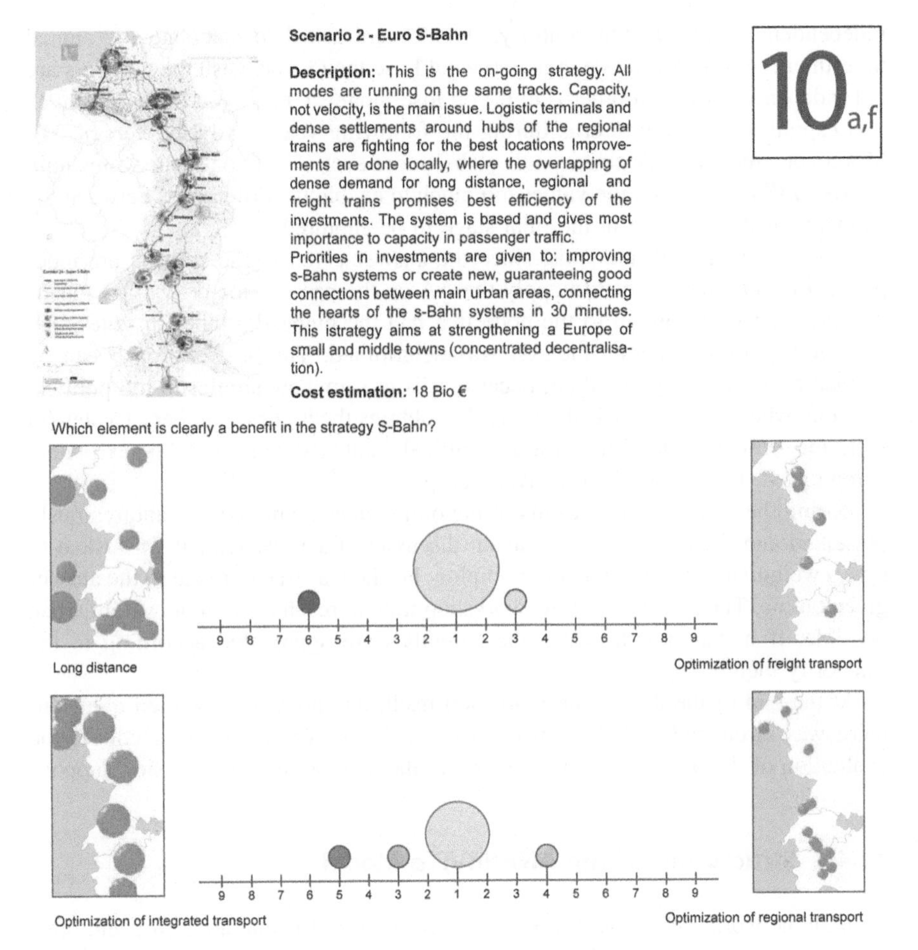

Scenario 2 - Euro S-Bahn
Description: This is the on-going strategy. All modes are running on the same tracks. Capacity, not velocity, is the main issue. Logistic terminals and dense settlements around hubs of the regional trains are fighting for the best locations Improvements are done locally, where the overlapping of dense demand for long distance, regional and freight trains promises best efficiency of the investments. The system is based and gives most importance to capacity in passenger traffic.
Priorities in investments are given to: improving s-Bahn systems or create new, guaranteeing good connections between main urban areas, connecting the hearts of the s-Bahn systems in 30 minutes. This istrategy aims at strengthening a Europe of small and middle towns (concentrated decentralisation).
Cost estimation: 18 Bio €
10a,f
Which element is clearly a benefit in the strategy S-Bahn?
9 8 7 6 5 4 3 2 1 2 3 4 5 6 7 8 9
Long distance
Optimization of freight transport
9 8 7 6 5 4 3 2 1 2 3 4 5 6 7 8 9
Optimization of integrated transport
Optimization of regional transport


**Fig. 12.1** Diagram showing the results from the online survey in relation to the question on the most beneficial strategy for scenario 2

comparison between two different strategies in relation to a scenario. As a result, most of answers show an indifference between the strategies, while the preference for a scenario is very light in term of number of preference. Therefore, the question was deemed as not answered and it was re-proposed to all the actors during the workshop.

After having collected all the responses of the online questionnaire, the weights were aggregated. In the literature many methods have been proposed to approach the aggregation. The most widespread methods are the Geometric Average (GA) and the Arithmetic Average (AA). The literature [3, 4] indicates the GA as the "evolution" of the AA but this does not mean that one is better than the other. It depends on the context of application. For example, if asked to find the class average of students' test scores, an AA would be used because each test score is an independent event. On the contrary, if one were asked to calculate the annual investment return of your savings you would use the GA because the numbers are not independent of each other (i.e. if you lose money 1 year, you have that much less capital to generate returns during the following years, and vice versa) [34]. Moreover, since the GA gives a null global score even if only one criterion is null, the risk will be to flatten the values too much to enable the differences between the elements of the decision in the final stage to be captured.

After applying both methods, since the answer given in the surveys are independent events, we decided to apply the AA on the basis of majority. This means that preference was given to the node that had the highest number of votes, and then among these weights the AA was determined. This last approach can be defined as a "majority" method, because it is in somehow similar to the political election, where the winner is the party that obtains the highest number of votes (in fact, the voting system in various political contexts worldwide is extremely complicated. For an in-depht analysis see [9].

During the workshop the results of the online questionnaire were anonymously presented and the questions were again discussed. Each participant was asked to give a weight and to explain his/her opinion; he/she was free to relate to the answer given online. The facilitator of the workshop tried to reach a common weight when possible. If it was not the case, the weights were aggregated according to the "majority method".

At the end of the day of the workshop itself, an anonymous written questionnaire was circulated and there was an open discussion, in order to collect the evaluation of the participants with regard to the methodology of decision support.

### 12.4.2 Structuring of the ANP/InViTo Model

A single network ANP model was developed in order to generate a free and open discussion among the stakeholders involved. The decision problem in exam has been divided into five clusters (namely, economic development, spatial development, rail operation, environment and logistics). Each cluster has been divided in turn into elements (or nodes) representing the specific aspects of the decision problem (Fig. 12.2).

Mention has to be made to the fact that in this application, the ANP is not used as a method to determine a priority list of the different alternatives in the decision problem. The ANP model here is applied as a structured procedure that is able to support the assessment of the identification of the considered key development factors in order to come to a decision. In this sense, it is a rather rare application of ANP technique because the obtained results are not an order of alternatives but a sorting of criteria [1, 10, 24].

Figure 12.3 represents the ANP model according to a single network structure. In order to improve the readability the Strategic Assessment Cards and the ANP questionnaire, a symbol has been associated to each element of the ANP network.

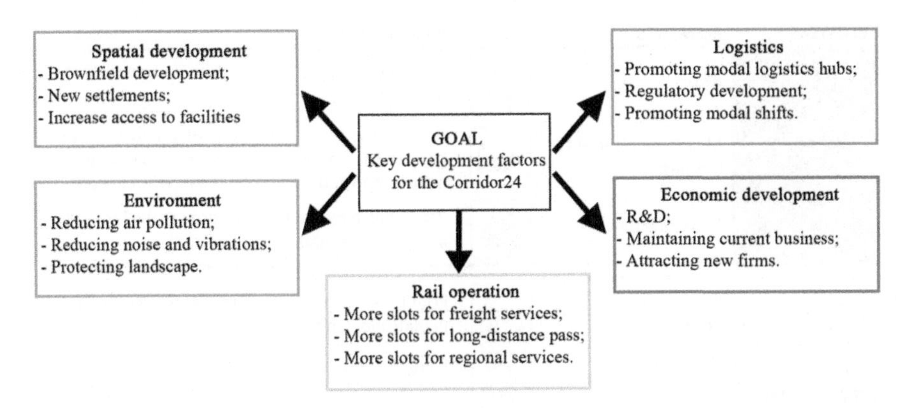


**Fig. 12.2** ANP single network model

After setting up the model it is necessary to develop the pairwise comparisons in order to establish the relative importance of the different elements, with respect to a certain component of the network. The comparison and evaluation phase is divided into two distinct levels: the cluster level, which is more strategic, and the node level, which is more specific and detailed. According to the methodology, in pairwise comparisons, the Saaty's absolute scale is used to compare any two elements [45]. The main eigenvector of each pairwise comparison matrix represents the synthesis of the numerical judgments established at each level of the network [43].

In the present application, the surveys containing all the pairwise comparisons coming from the ANP model was first submitted online to the stakeholders involved in the decision process.

The questions at the clusters lever were of the type:

Which aspects are more relevant for the strategy development of the Corridor24?

| Rail oper. | 9 | 8 | 7 | 6 | 5 | 4 | 3 | 2 | 1 | 2 | **3** | 4 | 5 | 6 | 7 | 8 | 9 | Economic |
|---|---|---|---|---|---|---|---|---|---|---|---|---|---|---|---|---|---|---|

The questions at the node level were of the type:

Which economic development aspect is more important for the strategy development for the Corridor24?

| Attracting new firms | 9 | 8 | 7 | 6 | 5 | 4 | 3 | 2 | 1 | 2 | 3 | **4** | 5 | 6 | 7 | 8 | 9 | R&D |
|---|---|---|---|---|---|---|---|---|---|---|---|---|---|---|---|---|---|---|

After collecting all the surveys, the weights were aggregated with the "majority method" (see Sect. 12.4.1).

In order to help the actors involved to understand the consequences of their choices, each ANP question was supported by a visualisation which aims at resuming the answers of actors themselves. For each answer, the visualisation should provide a feedback in real time, changing forms, sizes and colours on the basis of the values given by actors. In this way, actors can relate their answer with

| CLUSTERS | ELEMENTS | ICON | DESCRIPTION |
|---|---|---|---|
| Economic development | R&D | | Investments and specialization in research and development |
| | Maintaining current business | | Maintaining the current level of productions (e.g. businesses, industries, services) following the actual trend and strengthen the logistic clusters |
| | Attracting new firms | $ | Promoting new productions through the location of advanced and innovative firms |
| Spatial development | Brownfield development | | Prioritizing the reuse of brownfields over greenfield in metropolitan regions |
| | New settlements | | Increasing the housing stock in the metropolitan regions |
| | Increase access to facilities | | Improving the access to public and private services |
| Rail operation | More slots for freight services | F | Prioritizing the flow of freight services; new investments should focus on this goal. |
| | More slots for long-distance passengers services | L | Prioritizing the flow of long-distance passenger services; new investments should focus on this goal. |
| | More slots for regional services | R | Prioritizing the flow of regional passenger services; new investments should focus on this goal. |
| Environment | Reducing air pollution | | Reduction in pollutant emission by shifting from road to rail |
| | Reducing noise and vibrations | | Minimization of the impact of noise and vibrations from rail traffic |
| | Protecting landscapes | | Minimization of the visual impact of the train infrastructure |
| Logistics | Promoting multi-modal logistics hubs | | Improving the capacity of logistics hubs through access to the networks. |
| | Regulatory development | | Regulating and organizational actions/developments to improve the efficiency of logistic chains |
| | Promoting modal shift (towards rail) | | Incentives and political agreements to support modal shift (towards rail) |

**Fig. 12.3** Clusters, nodes and symbols of the ANP network

a visually explained effect on the strategy for the corridor. Therefore, visualisation acts as a common table for sharing data and collecting the answers to the ANP questionnaire.

On the basis of previous workshops, a first attempt used symbolic, dynamic maps for each alternative strategy option (Fig. 12.4). These maps were built

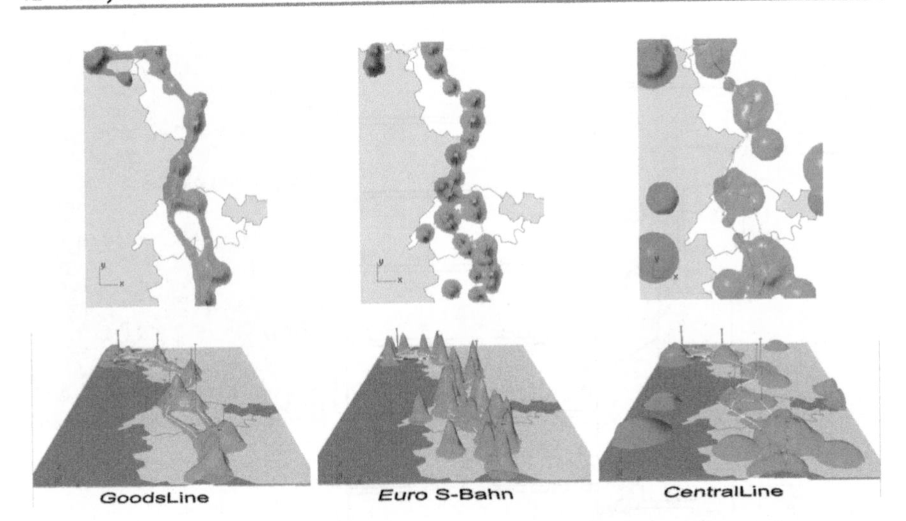


**Fig. 12.4** The first visualisation of the whole corridor: three-dimensional dynamic meshes which localise the costs and benefits along the whole corridor

by three-dimensional meshes which changed their location, height and footprint size on the basis of both actors' answers and ANP nodes. Different colours were used to highlight differences between Benefits and Costs, as described by the BC network used in the first study.

Nevertheless, the size of the area was vast and this kind of visualisation showed to be not effective in highlighting potential and critical key-factors along the corridor. The wide dimension required an innovative method in order to work as a *boundary object*. The new visualisation should deal with the complexity of planning issues, which should included spatial planning, economic and environmental aspects, transport and logistics interventions. Furthermore, it should provide information on the effect of actors' answers during the ANP questionnaire. Following this, a further study was developed to visualise the distribution along the corridor of the effects of each ANP node. Meanwhile, almost the same consideration on the size of project scale brought changes in the ANP structure too. In fact, the BC network has been changed in a single network.

The new study on visualisation focused on representing the single ANP network by offering a connection between spatial localisation and the expected effect of choices. Thus, a bi-dimensional geographical map of the whole corridor has been the basis for developing a new visualisation which could provide a direct correspondence between the nine sub-regions and the nodes of the ANP model (Fig. 12.5).

Also in this case, the visualisation resulted too complex to be easily understood in a multidisciplinary workshop. Therefore, a new simplification has been done in order to obtain an intuitive understanding of choices' effects on the whole corridor strategy. Histograms have been substituted by radial charts located on the main city of each sub-region (Fig. 12.6).

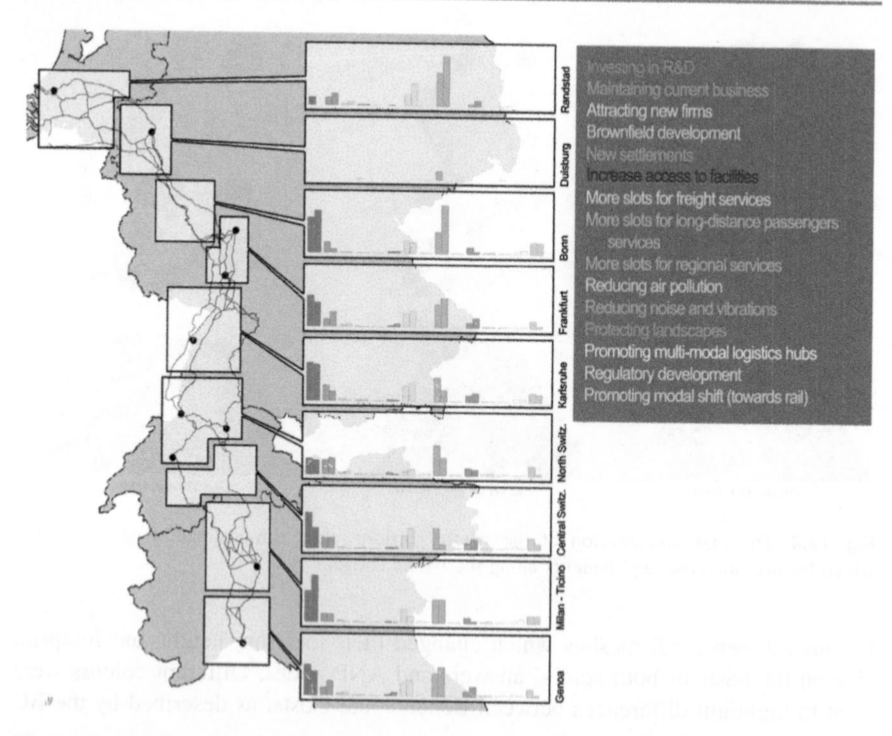


**Fig. 12.5** Visualisation which relates the strategic choices of workshop participants with their effects distributed along the different nine sub-regions of the Genoa–corridor

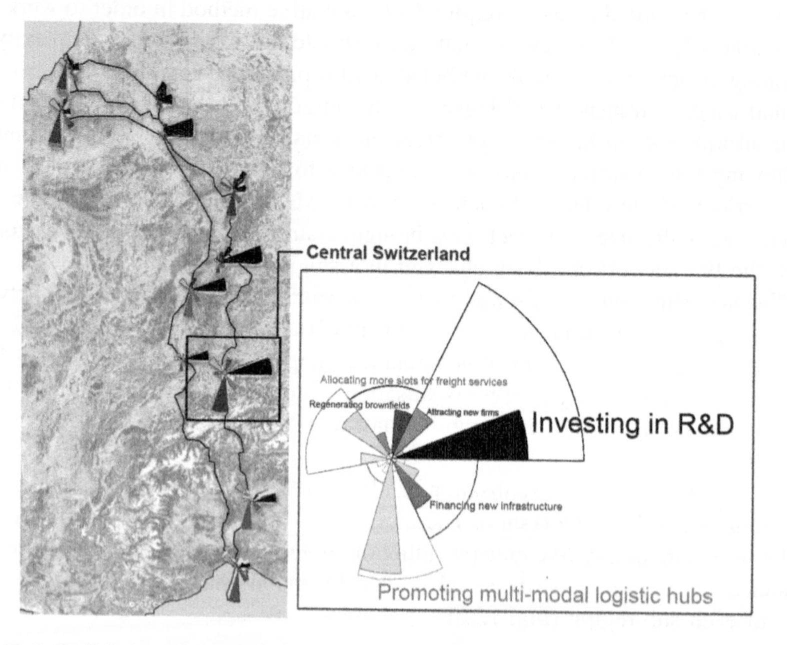


**Fig. 12.6** Radial chart distributed along the nine sub-region of the corridor

Radial charts aimed at highlighting, sub-region by sub-region, which aspect was more influencing in the development of the whole corridor. In fact, each radial chart was divided in circular sectors, one for each ANP node. The radius size of each sector was calculated on the value of weights given during the workshop. The interactive map changes its form with each question as to illustrate the answers of the participants [40].

Until this moment, the spatial localisation of answers was considered as essential information to understand the contribution of a common strategy to single sub-regions. In fact, the differences between the sub-regions could generate useful information to understand the distribution of advantages due to a strategy instead of other interventions.

The visualisation of the spatial distribution of ANP answers is a concept that has been abandoned, firstly for a lack of consistent data, secondly to avoid local preferences by the actors participating to the decision-making process.

As a result, this final visualisation has been essential to understand the influence of each ANP node in the strategy for the whole corridor. In particular, it provided a visual representation to the values of the weights given by participants, showing the changes in importance of single ANP nodes in real time.

### 12.4.3 Results of the ANP/InViTo Application

According to the ANP methodology, once all the pairwise comparison matrices has been compiled, it is possible to obtain the importance of the criteria and of the elements considered in the decision problem. In particular, while the priority of the criteria forms the cluster matrix, the priority of the elements forms the so-called unweighted supermatrix.

Table 12.2 represents the cluster matrix. The application of the cluster matrix to the unweighted supermatrix, as a cluster weight, provides the weighted supermatrix. This can be raised to a limiting power in order to obtain the limit supermatrix, in which all columns are identical and each column gives the global priority vector.

From the cluster matrix emerges that key development factors are related to the spatial development aspects of the affected areas (0, 27) followed by the logistics aspects (0, 24) and the economic aspects (0, 23).

At the element level, Fig. 12.7 represents the final priorities of the model.

The graph shows that the most important factors in the examined decision-making problem are: brown field development (0, 198) followed by the promotion of multi-modal logistic hubs (0, 147) and the reduction in noise and vibration (0, 145).

The results are in line with the results emerged during the comparisons at the cluster level.

The single chart of Fig. 12.8 constitutes a further simplification of the visualisation showed in Fig. 12.6. In fact, the radial charts located on the main cities have been delocalised and resumed in a single radial chart. The single chart has

**Table 12.2** Cluster matrix of the ANP network

| | Goal | Economic | Spatial | Rail oper. | Environ. | Logistics |
|---|---|---|---|---|---|---|
| Goal | 0.00 | 0.00 | 0.00 | 0.00 | 0.00 | 0.00 |
| Economic | 0.23 | 0.00 | 0.00 | 0.00 | 0.00 | 0.00 |
| Spatial | 0.27 | 0.00 | 0.00 | 0.00 | 0.00 | 0.00 |
| Rail oper. | 0.06 | 0.00 | 0.00 | 0.00 | 0.00 | 0.00 |
| Environ. | 0.20 | 0.00 | 0.00 | 0.00 | 0.00 | 0.00 |
| Logistics | 0.24 | 0.00 | 0.00 | 0.00 | 0.00 | 0.00 |

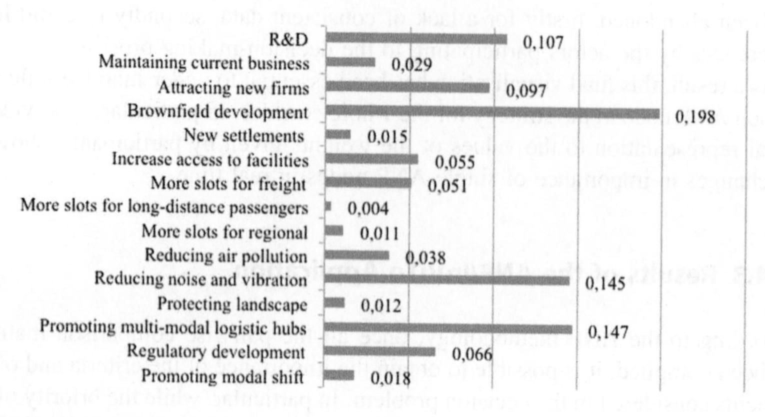


**Fig. 12.7** Final priorities of the elements of the model

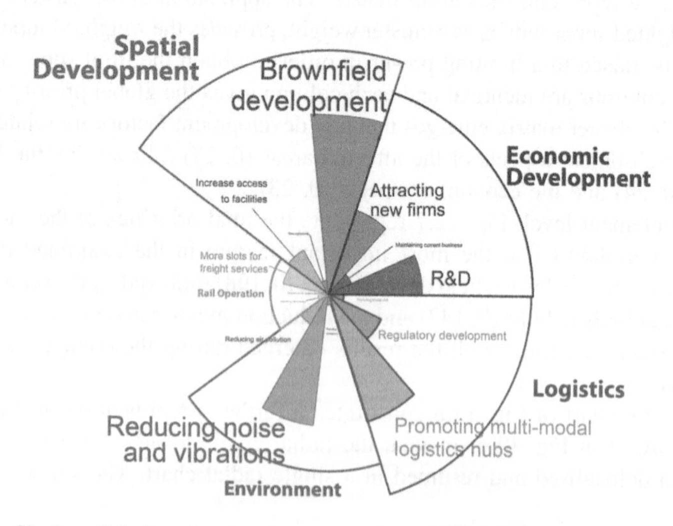


**Fig. 12.8** Single radial chart showing the results of the ANP/InViTo model

been proposed to the actors in combination with the single ANP network in order to better show the key factor for the corridor development. Moreover, the chart was fundamental to start the next step of the assessment (the strategic assessment).

### 12.4.4 Results of the Strategic Assessment

As aforementioned the collaborative assessment session has been designed in order to further deepen the ANP results by translating them into project components for the spatial strategy. The aim of the activity has therefore been to define the key elements of a possible common strategy. The strategy is produced in the form of recommendations addressing both the crucial interventions in the different regions and at the corridor's level.

As for the ANP, an intense preparation phase was needed prior to the workshop. The main difficulty in this case is to find the right balance between the need to providing a relevant amount of information to the participants, in order to allow their conscious contribution, and the limits of crucial resources (particularly time), which require the shared information to be easily comprehensible, manageable and transferrable. This means a great effort of simplification and classification, which must nevertheless be open to the new inputs coming from the discussion during the workshop. According to these features, the strategic assessment process can only be a revolving task which is constantly subjected to adjustments and reviewing by the designing participants. In this way two main goals are accomplished: a constant updating of the outcomes (which are always to be considered temporary) and, as a consequence, the strengthening of the participants' team into a permanent designing partnership.

From the contents point of view, the strategy's elements considered in the evaluation were of a different nature:

- Economic elements with specific reference to the future development scenarios for Europe. In this sense three different possible scenarios were designed: investing in research and development, maintaining the current level of production or promoting new productions through the location of advanced and innovative firms.
- Spatial development elements namely prioritizing the reuse of brownfields, increasing the housing stock in metropolitan regions and increase the access to facilities.
- Rail operation elements such as prioritizing the flow of freight service, of long-distance passenger service or of regional passengers services.
- Environmental elements with particular reference to the reduction in air pollution, the minimization in noise and vibrations or the minimization of the visual impact upon the landscape.
- Logistic elements such as improving the capacity of the logistics hubs, improving the efficiency of the logistic chains or promoting the modal shift.

For each region a first mapping of projects and problems has been conducted in order to provide a clear picture about a selection of issues which are considered to be critical. In particular the following items have been mapped:

- Problems and Projects in the railway network:
  - Insufficient capacity in the node.
  - Insufficient capacity of the line.
  - Upgrade of the existing infrastructure.
  - Upgrade of the regional services.
  - New high-speed line.
- Problems and Projects in the logistic network:
  - Hub with scarce development potential.
  - Hub with development potential.
- Settlement development and RE market:
  - Urban area with low demand and many development potentials.
  - Urban area with high demand and no development potentials.
  - Urban area with high demand and many development potentials.
- Landscapes threatened by noise and other impacts.

Different strategic development perspectives have been produced from this first elaboration which, in each case, focused on specific goals and features: e.g. on upgrading all long-distance connections between the biggest metropolitan areas or on increasing capacity for freight transport by adding new tracks or by-passes of the most congested nodes

For each region at least two alternative strategies have been developed. Alternative strategies are characterized by different policies of intervention (as presented also by the ANP's nodes) and connected sets of projects.

This information became material for the realization of a set of cards illustrating the different combinations of projects and expected outcomes. During the Strategic Assessment the participants were given the task to study and discuss the contents of the cards and to come to a joint strategy by assembling their preferred set of cards, also by adding new proposals. The core aim of the exercise was to then compare the results of the discussion with the outcome of the ANP in order to evaluate if and how abstract preferences would change when confronted with actual project proposals (Fig. 12.9).

As expected, the use of the strategies and of the cards has been very helpful in accompanying the discussion, providing a solid base for comparing options and the emergence of unsolved issues. In particular, the game setting enabled experts to put aside their role of advocates for their own regions and to give them the freedom to consider a wider perspective even if relying on little information. The effort helped also to verify the gap between abstract goals and feasible investments (technically and economically) and re-editing the ANP outcomes specifying the

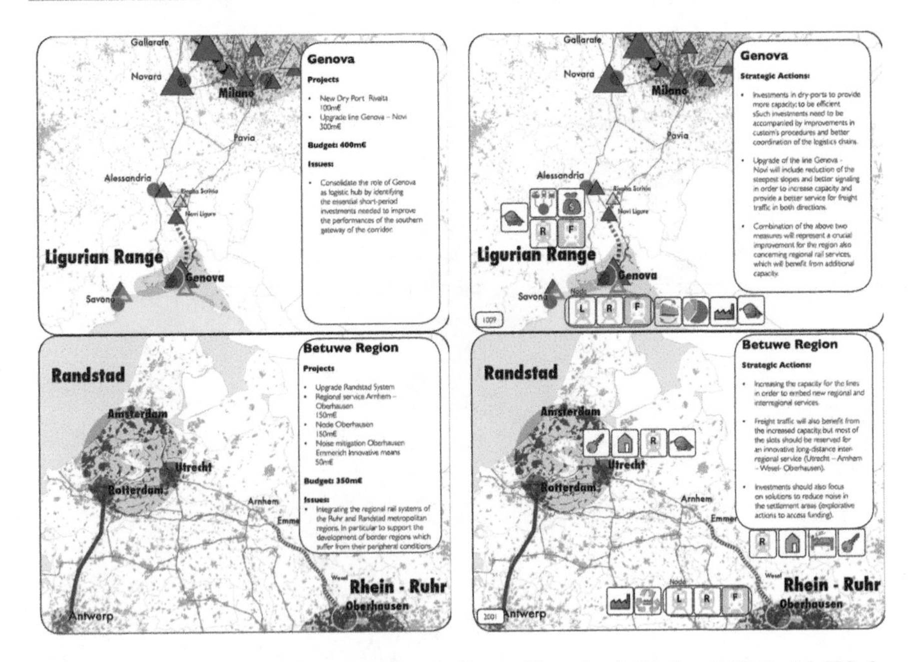


**Fig. 12.9** Strategic assessment: strategy cards (*Source* Tosoni and Günther, ETH Zurich IRL for INTERREG IVB NWE project CODE24)

key features of the elements (nodes) to be considered strategic. In some cases the participating experts felt the need to develop their own additional proposals by merging different goals (Fig. 12.10).

The result is therefore not one of the initial strategies envisaged, but a more complex action plan composed by intervention projects, guidelines for specific themes and the identification of 4 macro-regions where further development of the strategy is needed:

- The Mittelrhein: where an in-depth solution-design process should propose cost-benefit (efficient) solutions to the environmental problems.
- The Frankfurt-Mannheim region: where an efficient improvement of all networks could be achieved through flow separation.
- The Oberrhein region: where a better coordination between the Rhein network and the French network could be highly beneficial together with a smart solution for the gateway of Basel.
- The Ligurian range: where the empowerment of the port of Genova needs to be achieved through a gradual but intensive investment plan.

The final discussion not only consolidated these outcomes but also brought attention to the issue of future management of the strategy and its future upgrade, as well as the need for promoting continuous coordination and collaboration among the involved stakeholders.

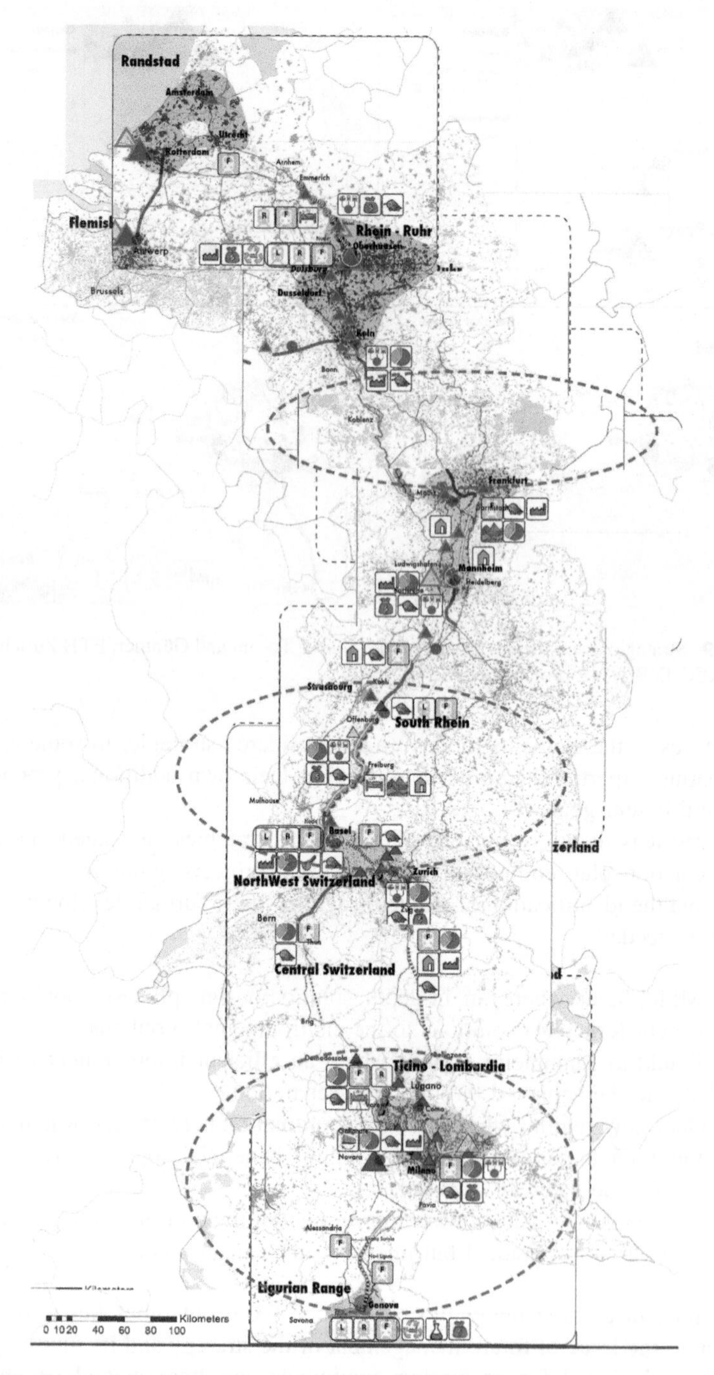


**Fig. 12.10** Corridor24 common strategy (*Source* Tosoni and Günther, ETH Zurich IRL for INTERREG IVB NWE project CODE24)

## 12.5 Conclusions

The aim of the Collaborative Assessment Procedure was to accompany the development of a shared position by the CODE24 project partnership regarding the most relevant issues affecting the future corridor's development. The assessment phase has therefore been an interactive process. It promoted an open and intensive discussion among the partners and other important stakeholders on the spatial and infrastructure development of the Rotterdam–Genoa corridor, in different regions and at interregional level. Through the procedure alternative development strategies for the whole Corridor24 have been explored.

The procedure has been designed in order to:

- Discuss the main components and principles of the common strategy with the participants, by comparing pairs of selected elements through the Analytic Network Process instrument (ANP).
- Jointly design the common strategies by choosing the best/optimal strategy among alternative options for each of the 9 subareas in which the corridor has been previously divided through the Strategic Assessment instrument.
- Develop and test visualisation tools (InViTo) able to support the decision making in real time.
- Discuss the implication (operational and political) of the chosen strategy and the further steps.

Through the Collaborative Assessment procedure and the challenging discussion between the experts it was possible to jointly outline a first draft of a common strategy composed by the following elements:

- An assessment of the main issues and problems for each region.
- An overview of strategic projects and their expected regional and interregional impacts as well as an indicative budget.
- A group of three regions where further explorative planning is needed to reach robust choices.
- The proposal of an operative mandate for the future EGTC (European Group of Territorial Cooperation) as operative platform for the maintenance of the partnership.

From the methodological point of view the experiment represented a groundbreaking activity. It provided the right environment for testing innovative instruments and tools, and collect important feedback on their performances. In particular, it seems that the combination of different inputs and modes of joint interaction is able to support a productive discussion also when dealing with extremely complex and challenging tasks. The workshops proved to be able to consolidate and enhance the collaboration between the participants, in spite of limited time and the difficulties arising from the different backgrounds of the stakeholders. The most significant sign of the action's success was the willingness of the involved partners to continue with the joint, collaborative work in the coming future.

## References


1. Abastante F, Lami IM (2013) An analytical model to evaluate a large scale urban design competition. Geoingegneria ambientale mineraria (GEAM) 139: 27–36
2. Ackoff R (1981) Creating the corporate future. Wiley, New York
3. Aczèl J, Roberts FS (1988) On the possible merging functions. Mathe Soc Sci 17:205–243
4. Aczèl J, Saaty TL (1983) Procedures for synthesizing ratio judgements. J Math Psychol 27:93–102
5. Albrechts L (2006) Problems and pitfalls in large-scale strategic planning. In: International congress "La Citta di Citta", Milano. 20 Feb 2006
6. Andersen DF, Richardson GP (1997) Scripts for group model building. Syst Dyn Interview 13(2):107–129
7. Balducci A (1991) Disegnare il futuro: il problema dell'efficacia nella pianificazione urbanistica. Il Mulino, Bologna
8. Black LJ, Andersen DF (2012) Using visual representations as boundary objects to resolve conflict in collaborative model-building approaches. Syst Res Behav Sci 29:194–208
9. Bouyssou D, Marchant T, Pirlot M, Perny P, Tsoukias A, Vincke P (2001) Evaluation and decision models: a critical perspective. Springer, Berlin
10. Bottero M, Lami IM (2010) Analytic network process and sustainable mobility: an application for the assessment of different scenarios. J Urbanism 3:275–293
11. Budge K, Beale C, Lynas E (2013) A Chaotic Intervention: Creativity and Peer Learning in Design Education. Int J Art Des Educ 32(2):146–156 (Blackwell Publishing Ltd)
12. Cascetta E (2009) Transportation system analysis: models and applications. Springer, New York
13. Crosta P (1995) La Politica del piano. Franco Angeli, Milano
14. Dematteis G, Governa F (2001) Urban form and governance: the new multi-centred urban pattern. In: Andersson H, Jorgensen G, Joye D, Osterndorf W (eds) Change and stability in urban Europe: form, quality and governance. Ashgate, Aldershot
15. Elster J (ed) (1998) Deliberative democracy. Cambridge University Press, Cambridge
16. Elster J, Aanund H (1990) Foundations of social choice theory. Cambridge University Press, Cambridge
17. Forester J (1999) The deliberative practitioner. Massachusettes Institute of Technology, Cambridge
18. Friend J, Hickling A (2005) Planning under pressure: the strategic choice approach, 3rd edn. Elsevier, Elsevier
19. Healey P (2002) Place, identity and local politics: analysing initiatives in deliberative governance. In: Hajer M, Wagenar H (eds) Deliberative politic analysis: understanding governance in the Network Society. Cambridge University Press, Cambridge
20. Healey P (2007) Urban complexity and spatial strategies. Routledge, Oxon
21. Innes J (1995) Planning theory's emerging paradigm: communicative action and interactive practice. J Plann Educ Res 14:138–196
22. Klein G, Klug N, Todes A (2012) Spatial planning, infrastructure and implementation: implications for planning school curricula. Town and Reg Plan 60
23. Innes J (1998) Information in communicative planning. J Am Plann Assoc 64(1):5263
24. Lami IM, Masala E, Pensa S (2011) Analytic network process (ANP) and visualization of spatial data: the use of dynamic maps in territorial transformation processes. Int J Anal Hierarchy Process (IJAHP) 3(2):92–106

25. Landry M, Malouin JL, Oral M (1983) Model validation in operations research. European Journal Oper Res 14:207–220
26. Latour B (2005) Reassembling the social. Oxford University Press, Oxford
27. Marchese U (2000) Lineamenti e problemi di economia dei trasporti. ECIG, Genova
28. Marshall S, Banister D (2007) Land use and transport: European research towards integrated policies. Elsevier, Oxford
29. Marttunen M, Mustajoki J, Dufva M, Karjalainen TP (2013) How to design and realize participation of stakeholders in MCDA processes? A framework for selecting an appropriate approach. EURO J Decis Process (forthcoming)
30. Maurer J (2005) Planerische Strategien und Taktiken. ARL: 758–764
31. McKenzei J, Winkelen CV (2011) Beyond words: visual metaphors that can demonstrate comprehension of KM as a paradoxical activity system. Syst Res Behav Sci 28(2):138–149
32. Mingers J, Brocklesby J (1997) Multimethodology: towards a framework for mixing methodologies. Omega 25(5):489–509
33. Mingers J, Rosenhead J (2004) Problem structuring methods in action. Eur J Oper Res 152:530–554
34. Mitchel DW (2004) More on spreads and non-arithmetic means. Mathe Gaz 88:142–144
35. Ostanello A (1997) Validation aspects of a prototype solution implementation to solve a complex MC problem. In: Clìmaco J (ed) Multi-criteria analysis. Springer-Verlag, Berlin, pp 61–74
36. Palermo PC, Ponzini D (2010) Spatial planning and urban development. Springer, Berlin
37. Pensa S, Masala E (2014) InViTo: an interactive visualisation tool to support spatial decision processes. In: Pinto NN, Tenedorio JA, Antunes AP, Roca J (eds) Technologies in urban and spatial planning: virtual cities and territories. IGI Global Book, Hershey (forthcoming)
38. Pensa S, Masala E, Lami IM (2013) Supporting planning processes by the use of dynamic visualization. In: Geertman S, Toppen F, Stillwell J (eds) Planning support systems for sustainable urban development 195. Springer, Berlin Heidelberg, pp 451–467
39. Pensa S, Masala E, Marina O (2013b) What if form follows function? The exploration of suitability in the city of Skopje. In: Mingucci R, Mourão Moura AC (eds) Disegnare Con 6(11):141–148
40. Pensa S, Masala E, Lami IM, Rosa A (2014) Seeing is knowing: data exploration as a support to planning. Civ Eng Spec Issue 167(CE5):3–8
41. Rosenhead J, Mingers J (eds) (2001) Rational analysis for a problematic word revised. Wiley, Chichester
42. Slowinski Roy B (2013) Question guiding the choice of a multicriteria decision aiding method. EURO J Decis Process 1:69–97
43. Saaty TL (1980) The analytic hierarchy process, planning, piority setting, resource allocation. McGraw-Hill, New York
44. Saaty TL (2005) Theory and applications of the analytic network process. RWS Publications, Pittsburgh
45. Saaty TL, Ozdemir MS (2005) The encyclicon: a dictionary of applications of decision-making with dependence and feedback based on the analytic network process. RWS Publications, Pittsburg
46. Schoemaker P (1998) Scenario planning: a tool for strategic thinking. In: Dyson R, OBrien F (eds) Strategic development: methods and models, pp 185–208. Wiley, Chichester
47. Scholl B (1995) Aktionsplanung. Zur Behandlung komplexer Schwerpunktaufgaben in der Raumplanung. ETH Zürich: vdf Verlag
48. Scholl B (2005) Strategische Planung. ARL, 1121–1129
49. Schönwandt W (2011) Grundriss der Raumordnung und Raumentwicklung. ARL
50. Schönwandt W, Voemanek K, Utz J, Grunau J, Hemberger C (2013) Solving complex problems. Jovis p 208. ISBN 9078-3-86859-244-3
51. Secchi B (2013) La città dei ricchi e la città dei poveri. Laterza, Roma

52. Smith B, MacGregor JT (1992) What is collaborative learning? In: National center on postsecondary teaching, learning, and assessment at Pennsylvania State University, vol 117(5), pp 10–30
53. Star S, Griesemer JR (1989) Institutional ecology, translations and boundary objects. Soc Stud Sci 19(3):387–420
54. Stone D (2002) Policy paradox. W.W. Norton & Company, New York
55. Topping KJ (2005) Trends in peer learning. Educ Psychol Int J Exp Educ Psychol 25(6):631–645
56. Tsoukiàs A (2007) From decision theory to decision aiding methodology. Eur J Oper Res 18:138–161
57. Vennix J (1996) Group model building: facilitating team learning using systems dynamics. Wiley, London
58. Weihrich H (1998) Daimler–Benzs move towards the next century with the TOWS matrix. In: Dyson R, OBrien F (eds) Strategic development: methods and models, pp 69–80. Wiley, Chichester

# 13 A Conjoint Analysis Exercise to Assess Quality Attributes in Freight Transport: Outcomes from a Survey Among Logistics Managers of Manufacturing Firms in North West Italy

Maurizio Arnone, Gerard de Jong, Tiziana Delmastro, Agnese Giverso and Domenico Inaudi

**Abstract**

This chapter reports the results of a conjoint analysis experiment performed in North West Italy to investigate shippers' sensitivity and preferences for four relevant freight transport attributes: transport cost, time, punctuality and risk of damage and loss. The empirical study and the subsequent model estimation provided interesting results: shippers evaluate quality attributes, such as punctuality and avoidance of risk of damage and loss more important than travel time savings and indicate a high willingness to pay to increase freight transport service reliability and safety.

M. Arnone · T. Delmastro (✉) · A. Giverso · D. Inaudi
SiTI—Higher Institute on Territorial Systems for Innovation,
via P.C. Boggio 61, 10138 Turin, Italy
e-mail: tiziana.delmastro@siti.polito.it

M. Arnone
e-mail: maurizio.arnone@siti.polito.it

A. Giverso
e-mail: agnese.giverso@gmail.com

D. Inaudi
e-mail: domenico.inaudi@polito.it

G. de Jong
Significance and ITS University of Leeds, Koninginnegracht 23,
2514 AB, The Hague, The Netherlands
e-mail: dejong@significance.nl

I. M. Lami (ed.), *Analytical Decision-Making Methods for Evaluating Sustainable Transport in European Corridors*, Sxi 11, DOI: 10.1007/978-3-319-04786-7_13,

## 13.1 Introduction

The objective of this study was to provide a quantitative assessment of relevant quality attributes of freight transport services by estimating parameters based on empirical research into shippers' evaluation of freight services. The basic idea to reach this aim was to simulate shippers' behaviour, modelling their decisions on hypothetical transport services. Conjoint analysis is currently used in transport economics to analyse this kind of transport choices.

Conjoint analysis is a stated preference-based technique allowing the assessment of different combinations of product features or attributes (such as the price), in order to understand the relative importance and the best combination of these elements in customers' opinion. It employs a carefully designed survey in which respondents are asked to rate, rank or choose (stated choice experiments) between different hypothetical product alternatives described by specific relevant attributes. Collected data are then usually combined with revealed preferences (based on actual choices) and analysed within the framework of discrete choice models.

The purpose behind conducting conjoint analysis experiments is to determine the relative influence of the design attributes upon the choices that are observed to be made by sampled respondents undertaking the experiment. In our study this translated into the desire of determining the relative importance and influence of the different freight transport attributes on shippers' decision making. Moreover, quantitative statistical estimations of freight attributes monetary values were obtained, giving indications about shippers' willingness to pay (WTP) or to accept (WTA) different transport services.

Logistics managers (shippers) of manufacturing firms located in North West Italy were interviewed, since they're people responsible for the decisions and the organization of goods shipments in their firms. 79 interviews were collected; the survey started in April 2011 and finished in June 2011.

Collecting information about freight transport attributes can be useful not only for researchers who use the statistical estimates to feed transport demand models, but also for the main stakeholders in the supply chain who can take advantage of firms' WTP/WTA knowledge in order to customize or propose new services. Also Public Authorities can use such information to address their decisions about investments and regulations for infrastructures and transport systems.

## 13.2 State of the Art

The conjoint analysis approach is an established procedure to collect stated preference information from respondents. In the freight transportation context this method has been used in several studies carried out throughout Europe. Most of the analysed literature makes use of stated choice experiments where respondents are asked to complete a number of choice tasks in which they have to select one or more alternatives defined by a set of attribute dimensions with different level values. The hypothetical alternatives proposed in the choice tasks often refer to the current situation (e.g. the actual shipment characteristics); in order to collect data about the companies' current shipments, but also about their location, and logistics organization, the stated preference survey is usually associated with a revealed preference questionnaire.

To investigate the most relevant attributes influencing logistics managers' decision-making, the most appropriate attribute levels to be included in the survey, as well as the types of choice models used to estimate parameters, a state of the art literature review on freight transport choice experiments was carried out. Some significant results, mostly referring to surveys carried out in Italy and Switzerland, are described in Fig. 13.1.

Although the studies adopt different approaches and methods, some common characteristics can be found: for instance, there's a wide consensus in considering reliability as one of the most relevant attributes for logistics managers' choices, while transport mode frequently proved to be a non-significant feature and researchers often chose to carry out within-mode experiments.

## 13.3 Survey Design

On the basis of the analysed literature, a survey made up of both a revealed and a stated preference part was expressly designed.

The revealed preference (RP) questionnaire aimed to collect general information about the interviewed companies and above all about their "benchmark shipment", a typical shipment that respondents were asked to describe, possibly both for incoming and outgoing goods, and to take as a reference in the following stated preference (SP) survey.

The stated preferences (SP) were collected through a choice experiment composed by 20 choice tasks where interviewees had to choose between two hypothetical transport alternatives described by different levels of the same attributes.

After the selection of the sample of firms in charge of organizing their own shipments, meetings with their logistics managers were arranged in order to conduct the survey face to face through computer assisted personal interviewing (CAPI) method.

### 13.3.1 Revealed Preferences

The Revealed Preference questionnaire was divided into three sections. The first one included general questions about company size and location: company name and headquarter address, NACE production sector (NACE, ATECO in Italy, is the statistical classification of economic activities in the European Community; Rev 1.1-2002 was used), average turnover, number of employees, location of production and distribution centres (Fig. 13.2).

The second group of questions was focused on the description of the "benchmark shipment" (Fig. 13.3). Depending on the type of shipments handled

| Author<br>Title | Survey Area | Respondents | Attributes | Attribute levels | Number of alternatives in each choice task | Labelled/ Unlabelled alternatives | Number of choice tasks | Current scenario | Estimation model | Main results |
|---|---|---|---|---|---|---|---|---|---|---|
| Kofteci S., Ergun M., Serpil Ay H.<br>Modelling freight transportation preferences: Conjont analysis for Turkish Region [7]<br>2010<br>Research paper | Turkish Region of Antalya | Logistics managers of 50 cement firms | Transport Mode | Road / Intermodal | 2 | Unlabelled | 18<br>In total the survey includes 33 questions divided in 4 groups:<br>- rating (5 questions)<br>- importance (5 questions)<br>- pairs (18 questions)<br>- calibration (5 questions) | no | MNL (Multinomial Logit Model) | Time reliability is the most important attribute for a freight transport service choice, followed by cost, time, transport mode. Damage and loss is not an important attribute for logistics managers' decisions. |
| | | | Transport Cost | -10% / -5% / Current cost / +5% / +10% | | | | | | |
| | | | Transport Time | - half day / Current trasnport time / + half day / + 1 day / + 2 days | | | | | | |
| | | | Time reliability (% of transport services arriving on time per year) | All shipments on time / 80% / 60% | | | | | | |
| | | | Damage and loss (% of shipments damaged or lost) | None / ~ 5% / ~ 10% | | | | | | |
| Masiero L., Hensher D.A.<br>Analyzing loss aversion and diminishing sensitivity in a freight trasport stated choice experiment [10]<br>2010<br>Research paper | Ticino Region (Switzerland) | Logistics managers of 27 manifacturing firms | Transport Cost (CHF) | -10% / -5% / Current cost / +5% / +10% | 3 | Labelled (by transport mode: road, combined transport, piggyback) | 15 | yes | Mixed Logit | A symmetric and two asymmetric (taking separately into account increasing and decreasing attributes) models were estimated. The result suggest a significant improvement in model fit godness when preferences are asymmentric. The relevance of punctuality in freight transport is confirmed. This attribute is followed by time and cost. When asymmetries are considered the WTP for time saving and punctuality decreases while the WTA is higher: this disparity supports the loss aversion assumption, that is losses are valued more highly than gains. |
| | | | Transport Time | -10% / -5% / Current transport time / +5% / +10% | | | | | | |
| | | | Punctuality (% of transport services arriving on time per year) | 100% / 98% / 96% | | | | | | |
| Masiero L., Maggi R.<br>Accounting for WTP/WTA discrepancy in discrete choice models: discussion of policy implications based on 2 freight transport stated choice experiments [11]<br>2010<br>Research paper | Switzerland | Swiss logistics managers of 35 firms operating in the food and wholesale sector - Dataset 2003 | Transport Cost (CHF) | -40% / -20% / Current cost / +20% / +40% | 2 | Unlabelled | 20 | no | Mixed Logit | The results show that the asymmetric model (reference dependent specification considering attribute gains and losses) outperform the symmetric specification. Loss aversion has been registered for all attributes investigated in the analysis leading to a significant WTA/WTP discrepancy. Punctuality still remains a crucial factor, specially when logistics managers are faced with a reduction of this attribute. |
| | | | Transport Time | -40% / -20% / Current transport time / +20% / +40% | | | | | | |
| | | | Transport punctuality (% of transport services arriving on time per year) | 100% / 98% / 96% | | | | | | |
| | | | Damages (% of damaged shipments) | 6% / 4% / 2% | | | | | | |
| | Switzerland | Swiss logistics managers of 27 medium and large manufacturing firms - Dataset 2008 | Transport Cost (CHF) | -10% / -5% / Current cost / +5% / +10% | 3 | Labelled (by transport mode: road, combined transport, piggyback) | 15 | yes | | |
| | | | Transport Time | -40% / -20% / Current transport time / +20% / +40% | | | | | | |
| | | | Transport punctuality (% of transport services arriving on time per year) | 100% / 98% / 96% | | | | | | |
| Maggi R., Rudel R.<br>The value of quality attributes in freight transport: evidence from an SP-experiment in Switzerland [8]<br>2008<br>Research paper | Switzerland | 35 shippers of medium and large companies of the food and wholesale sector in Swiss firms. | Transport cost | - 20% / -10% / Current cost / +10% / +20% | 2 | Unlabelled | 20 | no | MNL (Multinomial Logit Model) | Two models have been estimated: 1. with the transport price in linear form 2. with a distance-dependent transport price. Logistics managers evaluate avoidance of damages and punctuality more important than travel time savings. The negative impact of the price attribute increses with distance. |
| | | | Transport time | -20% / -10% / Current transport time / +10% / +20% | | | | | | |
| | | | Frequency of delay | 98% / 95% / 90% | | | | | | |
| | | | Frequency of damage | 98% / 96% / 94% | | | | | | |

**Fig. 13.1** Literature review on freight transport choice experiments: main results

| Author<br>Title | Survey Area | Respondents | Attributes | Attribute levels | Number of alternatives in each choice task | Labelled/ Unlabelled alternatives | Number of choice tasks | Current scenario | Estimation model | Main results |
|---|---|---|---|---|---|---|---|---|---|---|
| Danielis R., Marcucci E., Rotaris L<br><br>Logistics managers' stated preferences for freight service attributes [2]<br><br>2005<br>Research paper | Friuli Venezia Giulia and Marche | Logistics managers of 65 small or medium size manufacturing firms (35 in Friuli Venezia Giulia, 30 in Marche) | Transport cost | -10% / -5% / Current cost / +5% / +10% | 2 (with a 9 point scale) | Unlabelled | n.d. | no | Logit and Ordered Probit Model | Estimates indicate a strong preference for quality attributes over costs. They indicate a high willingnes to pay for quality in freight transport services, especially for reliability and safety. Results denote a high aversion to risk of delay: an hour of unexpected delay is valued about 50% more than an hour of expected travel time. |
| | | | Transport time | Current transport time / + 1 day / +3 days / +5 days | | | | | | |
| | | | Punctuality (risk of delay) | No risk / Half day of delay / 1 day of delay / 3 day delay | | | | | | |
| | | | Risk of damage and loss (% of shipment value damaged or lost) | No risk / 5% of shipment value / 10% of shipment value | | | | | | |
| Zotti J., Danielis R.<br><br>Freight transport demand in the mechanics' sector of Friuli Venezia Giulia: the choice between intermodal and road transport [17]<br><br>2004<br>Research paper | Friuli Venezia Giulia | 30 companies of the mechanics sector in Friuli Venezia Giulia | Transport cost | -15% / -10% / -5% / Current cost / +5% / +10% / +15% | 3 | Unlabelled | 15 | yes | Multinomial Logit/Mixed Logit, Latent Class Model | The results of this survey show that transportation mode doesn't represent a significant choice variable, while attributes related to the quality of service are even more important than the cost attribute. |
| | | | Transport time | - half day / Current transport time / + half day / + 1 day / + 2 days | | | | | | |
| | | | Time reliability (% of transport service arriving on time per year) | 100% / 85% / 70% | | | | | | |
| | | | Damages and losses (% of shipments damaged or lost) | 0% / 5% / 10% / 20% | | | | | | |
| | | | Transport mode | Intermodal / Road | | | | | | |
| | | | Frequency | High / Low | | | | | | |
| | | | Flexibility | High / Low | | | | | | |
| Beuthe M., Bouffioux C., De Maeyer J., Santamaria G., Vandresse M., Vandaele E., Witlox F.<br><br>A Multi-Criteria Methodology for Stated Preferences Among Freight Transport Alternatives [1]<br><br>2002<br>Research Paper<br>(preliminary output of a research) | Belgium | Freight transport managers (around 125 firms) | Cost (CHF per transport service) time (hours per transport service) Loss, frequency, reliability, flexibility | - 20% / -10% / Current / +10% / +20% | No choice but ranking of 25 alternatives | Unlabelled | n.d. | yes | Utility additive (UTA) multicriteria method | Preliminary outputs of the research are presented. Transport cost is the most important factor followed by reliability but with a much lower weight. The other factors take some importance in a few cases accrding to the transport circumstances. The non-cost quality attributes, taken together weight almost as much as the cost. |
| Fowkes T., Shinghal N.<br><br>Freight mode choice and adaptive stated preferences [14]<br><br>2002<br>Research paper | India - Delhi-Bombay Corridor | 32 firms from six different production sectors: 7 pertained to export traffic and 25 to domestic traffic | Transport Cost (door to door shipments) | The Leeds Adaptive Stated Preference Software was used. LASP uses a four column format, with the initial attribute levels being based on the data about the currently used mode. The attribute levels for subsequent iterations are modified on the basis of the ratings given in the immediatly precedig iterations. | 4 | Labelled (existing road service, new road service, container service, rail service - speed link) | 9 | yes | Logit | The survey suggests a dislike for through rail services. The frequency of service appears to be an important factor in mode choice especially for the manufactured goods sector. The reliability of transit times appears to be very important for exporters and also for the autopart sectors. The results suggest that intermodal services can be viable in India for high value and finished goods, but these would need high frequency, reliable and fast services. Rail can be a viable service for the bulk good sector. |
| | | | Door to door transport time (with increments of 1/3 of a working day: morning delivery, afternoon delivery, evening delivery) | | | | | | | |
| | | | Punctuality (% of consignments arriving on time) | | | | | | | |
| | | | Frequency of service (daily, tri-weekely and weekely) | | | | | | | |

**Fig. 13.1** (continued)

**Company size and location**

1. Company name:
2. Contact name:
3. Headquarter address:
4. Production sector (NACE classification):
   - ☐ Manufacture of food products; beverages and tobacco
   - ☐ Manufacture of textiles and textile products
   - ☐ Manufacture of leather and leather products
   - ☐ Manufacture of wood and wood products
   - ☐ Manufacture of pulp, paper and paper products; publishing and printing
   - ☐ Manufacture of coke, refined petroleum products and nuclear fuel
   - ☐ Manufacture of chemicals, chemical products and man-made fibres
   - ☐ Manufacture of rubber and plastic products
   - ☐ Manufacture of other non-metallic mineral products
   - ☐ Manufacture of basic metals and fabricated metal products
   - ☐ Manufacture of machinery and equipment n.e.c
   - ☐ Manufacture of electrical and optical equipment
   - ☐ Manufacture of transport equipment
   - ☐ Manufacturing n.e.c.
5. Average revenue per year:
   - ☐ Less than 250.000 €
   - ☐ 250.000 – 500.000 €
   - ☐ 500.000 – 2.000.000 €
   - ☐ 2.000.000 – 10.000.000 €
   - ☐ 10.000.000 – 50.000.000 €
   - ☐ More than 50.000.000 €
6. Number of employees:

   ☐ 20 – 49 ☐ 50 – 99 ☐ 100 – 249 ☐ 250 – 499 ☐ 500 or more
7. Number and location of production plants (P) and/or distribution centres (D) - please insert number in the appropriate box:

| P | D | P/D | |
|---|---|---|---|
| | | | Piemonte |
| | | | Lombardia |
| | | | Liguria |
| | | | Valle d'Aosta |
| | | | North-East Italy |
| | | | Central Italy |
| | | | Southern Italy |

| P | D | P/D | |
|---|---|---|---|
| | | | Switzerland |
| | | | France |
| | | | Germany |
| | | | Netherland |
| | | | Belgium |
| | | | Other European countries |
| | | | Other non European countries |

**Fig. 13.2** First part of the RP questionnaire: "Company size and location"

(incoming goods, outgoing goods or both) logistics managers were asked to choose and describe a shipment:

- Important for their company (i.e. in terms of frequency or goods volume).
- Departing or arriving from/to a production or distribution unit located in North West Italy.
- Possibly carried out along Corridor24 Genoa-Rotterdam.

The chosen shipments were then used as a benchmark for the description of different transport alternatives in the following hypothetical choice experiment

**Description of the "BENCHMARK SHIPMENT"**

8. **Type** of goods:

| | INCOMING GOODS | OUTGOING GOODS |
|---|---|---|
| Description | | |
| Commodity classification –NST2007– see annex | | |
| 9. **Weight** of goods (tonnes): | | |
| 10. **Value** of goods (€): | | |
| 11. **Origin** (town/Country): | | |
| 12. **Destination** (town/Country): | | |

13. **Transport mode**:

| INCOMING GOODS | OUTGOING GOODS |
|---|---|
| ☐ Road | ☐ Road |
| ☐ Rail | ☐ Rail |
| ☐ Road + rail | ☐ Road + rail |
| ☐ Road + inland waterways | ☐ Road + inland waterways |
| ☐ Road + air | ☐ Road + air |
| ☐ Road + sea | ☐ Road + sea |
| ☐ Rail + sea | ☐ Rail + sea |
| ☐ Other:__________ | ☐ Other:__________ |

| | INCOMING GOODS | OUTGOING GOODS |
|---|---|---|
| 14. **Preferential path** (alpine crossing/route used): | | |
| 15. **Average distance** (km): | | |
| 16. **Cost** (€): | | |
| 17. **Transport time** (hours – door to door): | | |
| 18. **Number of shipments** of the benchmark type per year: | | |
| 19. **Acceptable delay** – max % of additional transport time that you would accept in order to consider the shipment still on time: | | |
| 20. **Delay** – % of delayed arrivals (exceeding the acceptable delay) per year: | | |
| 21. **Risk of damage and loss** – % of shipments per year suffering damage and loss which cause the damaged products to be replaced: | | |

22. **Load type**:

| INCOMING GOODS | OUTGOING GOODS |
|---|---|
| ☐ Bulk (solid or liquid) | ☐ Bulk (solid or liquid) |
| ☐ Unitized goods (container, swap bodies) | ☐ Unitized goods (container, swap bodies) |
| ☐ General cargo (parcels, pallets, coils, ecc.) | ☐ General cargo (parcels, pallets, coils, ecc.) |
| ☐ Other:__________ | ☐ Other:__________ |

23. **Responsible for the shipment**:

| INCOMING GOODS | OUTGOING GOODS |
|---|---|
| ☐ Own account | ☐ Own account |
| ☐ Third parties account | ☐ Third parties account |

24. Please **rank** from most important (1) to least important (5) the following freight transport attributes:

| 1 | 2 | 3 | 4 | 5 | | 1 | 2 | 3 | 4 | 5 |
|---|---|---|---|---|---|---|---|---|---|---|
| ☐ | ☐ | ☐ | ☐ | ☐ | Shipment cost | ☐ | ☐ | ☐ | ☐ | ☐ |
| ☐ | ☐ | ☐ | ☐ | ☐ | Transport time | ☐ | ☐ | ☐ | ☐ | ☐ |
| ☐ | ☐ | ☐ | ☐ | ☐ | Punctuality | ☐ | ☐ | ☐ | ☐ | ☐ |
| ☐ | ☐ | ☐ | ☐ | ☐ | Risk of damage and loss | ☐ | ☐ | ☐ | ☐ | ☐ |
| ☐ | ☐ | ☐ | ☐ | ☐ | Transport mode | ☐ | ☐ | ☐ | ☐ | ☐ |

**Fig. 13.3** Second part of the RP questionnaire: "Benchmark shipment description"

(stated preference experiment). Referring to the benchmark shipment respondents were asked questions about the type (NST2007 commodity classification), weight and value of transported goods, origin and destination of the shipment, transport mode (multiple choice among road, rail or different types of combined transport), preferential path, if known (e.g. alpine crossing or corridor used), and average distance covered in kilometres. If distance data were missing the information about shipment origin, destination, route and transport mode allowed us to calculate on maps the covered kilometres.

Other questions were related to the transport cost and time (thinking about the door to door shipment), number of shipments of the benchmark type per year, acceptable delay (additional transport time accepted to consider the benchmark shipment still on time), number of late arrivals per year (in percentage points referring to the acceptable delay), number of shipments suffering damage or loss per year (in percentage points). The last two attributes, together with transport cost and time, represent the reference factors used to depict the hypothetical transport alternatives in the SP experiment.

The last questions in this part were about load type (bulks, unitized goods, general cargo, etc.) and shipment provider (own account/ third parties account); finally logistics managers were asked to rank 5 freight transport attributes (transport cost, time, punctuality, risk of damage and loss, mode) using a rating scale from 1 (most important) to 5 (least important). This exercise showed once again, according to the literature, that after controlling for other factors transport mode is (among the selected attributes) one of the least important factors in logistics managers' opinion.

The last part of the questionnaire investigated company transport and logistics as it can be seen in Fig. 13.4 (only the part referred to incoming goods is shown, the same questions were asked also for outgoing goods shipments). Information about the modal split was collected asking the modal share (in percentage values) of both goods volumes (tonnes) transported and number of shipments per year.

Interviewees were then asked to list different types of imported/exported goods, indicating if they're in charge of organizing their shipment, the location of main suppliers/customers and to describe their inventory management method.

The RP questionnaire fixed structure was read by an Optical Character Recognition (OCR) software; through this method, answers were automatically collected in a dedicated database where each row represents an interview and each column includes a specific answer.

### 13.3.2 Stated Preferences

The second part of the survey consists of a stated choice experiment.

Stated choice studies typically consist of numerous respondents being asked to complete a number of choice tasks in which they have to select one or more alternatives among a finite set of alternatives. In each task, the alternatives are typically defined on a number of different attribute dimensions, each of which are further described by pre-specified levels drawn from some underlying *experimental design*.

**Company transport and logistics**

INCOMING goods:

25. Total volume per year and modal split for incoming goods (% on total transported volume):

| | Road (%) | Rail (%) | Road + Rail (%) | Road + inland waterways (%) | Road + Air (%) | Road + Sea (%) | Rail + Sea (%) | Others ______ (%) | TOTAL |
|---|---|---|---|---|---|---|---|---|---|
| Tonnes/year | | | | | | | | | 100 |
| N. shipments/year | | | | | | | | | 100 |

26. Type of incoming goods (NST2007 classification – see appendix) – Please, check the box (S) if you organize the shipment:

| S | Description | NST 2007 |
|---|---|---|
| | | |
| | | |
| | | |

| S | Description | NST 2007 |
|---|---|---|
| | | |
| | | |
| | | |

27. Location of main suppliers (max. 5):

- ☐ Piemonte
- ☐ Lombardia
- ☐ Liguria
- ☐ Valle d'Aosta
- ☐ North-East Italy
- ☐ Central Italy
- ☐ Southern Italy
- ☐ Switzerland
- ☐ France
- ☐ Germany
- ☐ Netherland
- ☐ Belgium
- ☐ Other European countries
- ☐ Other non European countries

28. Inventory management for incoming goods:
(E.g. Minimal or non-existing warehousing (JIT), stocks greater than or equal to reorder point, replenishing on real demand, replenishing on planned demand, replenishing on production scheduling, etc.)

______________________________________________

______________________________________________

**Fig. 13.4** Third part of the RP questionnaire: "Company transport and logistics"

The allocation of the attribute levels within the experimental design may impact upon the independent assessment of each attribute contribution to the observed choices as well as upon the statistical power of the experiment as its ability to detect statistical relationships that may exist within the data.

Conceptually, an experimental design may be viewed as a matrix whose values represent the attribute levels that will be used in the stated choice survey, whereas matrix rows and columns represent the choice situations (tasks), attributes and alternatives (groups of columns) of the experiment.

In our study shippers were asked several times to choose between two hypothetical freight transport alternatives (binary choice tasks were chosen to simplify

**Table 13.1** List of attributes and levels used in the choice experiment

| Attributes | Levels |
|---|---|
| Transport cost | −15 % |
| | −5 % |
| | Current cost |
| | +5 % |
| | +15 % |
| Transport time | −15 % |
| | −10 % |
| | Current time |
| | +10 % |
| | +25 % |
| Delay—percentage of shipments per year arriving late | 0 % |
| | 5 % |
| | 10 % |
| | 15 % |
| | 20 % |
| Risk of damage and/or loss—percentage of shipments per year suffering damage and/or loss wich cause the damaged goods to be replaced | 0.01 % |
| | 0.10 % |
| | 1.00 % |
| | 3.00 % |
| | 6.00 % |

as much as possible respondents' choices, also the status quo or none alternative was excluded with the aim of loosing as little information as possible) in order to assess the most important attributes affecting their decision making, when they organize a freight shipment.

There are many significant freight transport factors influencing companies' shipments: transport cost, special offers or discounts, transport time, service frequency, time of departure, punctuality, reliability, capacity to carry out urgent deliveries or to organize special shipments, origin and destination proximity, etc. Some of them are financial attributes, others depend on transport time, others are related to transport service, route or shippers' characteristics.

According to the literature (similar studies and surveys were carried out in different Italian regions and in other European countries as it can be seen from Fig 13.1) and with the aim of using a reasonably small number of attributes, reducing respondents' cognitive burden, four attributes were selected as probably the most important to describe the different transport services:

1. Transport cost.
2. Transport time.
3. Transport delay.
4. Risk of damage and loss.

Transport mode wasn't considered since numerous studies demonstrated that this attribute is not very significant: firms are not concerned about the means used to ship their products provided that the desired levels of other transport attributes are satisfied. In recent years, with transport outsourcing diffusion, transport demand has become a transport service request, as Danielis [4] states; this means that manufacturing companies nowadays choose services combining satisfactory cost, time and quality levels. These attributes involve and affect the mode choice, but transport mode is directly selected only by forwarders or logistics operators supplying transport services to the firms. So manufacturing companies can have an influence on transport mode choice, but normally don't choose it directly and specifically.

The set of levels for each attribute was chosen in order to provide a realistic enough description of the proposed hypothetical transport alternatives without exceeding the values considered by too much (this could result in respondents taking the survey no longer seriously).

On the basis of experts' suggestions and of internal test results, the final levels chosen for the survey design were the ones listed in Table 13.1 for each selected attribute.

Since different transport services can show very different values in terms of the chosen attributes, levels were defined in percentage points. Relative levels (percentage variation with reference to the current values) were chosen for transport cost and time, while absolute levels were used for delay and risk of damage and loss. In order to understand the identified level values, the last two attributes need to be interpreted as the number of trips/year arriving late (with respect to the revealed acceptable delay) and the number of shipments/year suffering damage and/or loss which require that the damaged goods will be replaced.

The number of selected attributes and levels would generate $5^4 = 625$ possible transport service alternatives, this means that the use of a full factorial design, in which all possible attribute level combinations are enumerated allowing the treatment of attributes interactions, would have been impossible since too many choice tasks were needed.

In order to reduce the design size, a fractional factorial design (that uses a carefully selected subset of choice tasks from the full factorial design considering principal effects only) was chosen and through the Ngene software, a sequential orthogonal experimental design was generated. Orthogonality provides that all attributes are statistically independent of one another (i.e. the correlation between each two attributes is zero), thus ensuring that the attribute combinations presented are varied independently from one another and that the effect of each attribute on the responses is more easily identified. Sequentiality means that the design is constructed by generating an orthogonal design for a single alternative, and then, starting from this profile, orthogonal designs for the other alternatives are constructed: in this way smaller designs can be found more easily, but they are only orthogonal within each alternative, not across alternatives.

The generated design was attribute level balanced (i.e. for each attribute, all levels appeared an equal number of times in the questionnaire thus avoiding that more frequently occurring levels were measured with greater precision than less frequent ones) and originally included 25 binary choice tasks. 6 choice tasks presented a dominant alternative, that is an alternative clearly superior in utility to the other one: a choice of a dominant alternative is less informative for identifying respondents' preferences than a more thoughtful trade-off among alternatives, where there isn't an obvious choice, moreover respondents could be annoyed by a high degree of dominance which requires trivial answers. For these reasons 5 of the 6 choice tasks were removed and a final design of 20 binary choice tasks, that sacrifices a modest degree of orthogonality but produces better overall results, was used.

The stated choice experiment was administered through face to face interviews with the help of a personal computer and of a proper input mask created in Excel VBA. Respondents were asked to state their preferences about the proposed alternatives always referring to their benchmark shipment, this means that in each task, they had to choose which of the two optional transport services they would use for the typical freight shipment just described in the revealed preference (RP) questionnaire.

Firstly, interviewees were asked to insert their company name, date of compilation and type of benchmark shipment they were referring to (incoming or outgoing goods). Then they had to fill in a second window with the selected attribute (cost, time, delay, risk of damage/loss) values just defined for their benchmark shipment; using these values as a reference, the 20 choice tasks were then proposed both with percentage levels and with absolute levels which, for the first two attributes, were calculated in real time on the basis of the inserted benchmark shipment values, as shown, for example, in Fig. 13.5.

The choices stated by the interviewed shippers were automatically organized in a database, used for the model estimation. For each respondent, the chosen alternatives were listed next to the attribute levels describing the two hypothetical transport services proposed in each choice situation.

## 13.4 Choice Model

### 13.4.1 Model Estimation

A **binary logit model** was defined [the same model as the multinomial logit (MNL) model described in Chap. 8 on discrete choice, but with only two choice alternatives].

In our case the systematic utilities associated to each alternative are defined as linear combinations of the previously selected freight transport attributes:

$$V_1 = ASC + \beta_c c_1 + \beta_t t_1 + \beta_p punct_1 + \beta_r risk_1$$
$$V_2 = \beta_c c_2 + \beta_t t_2 + \beta_p punct_2 + \beta_r risk_2$$

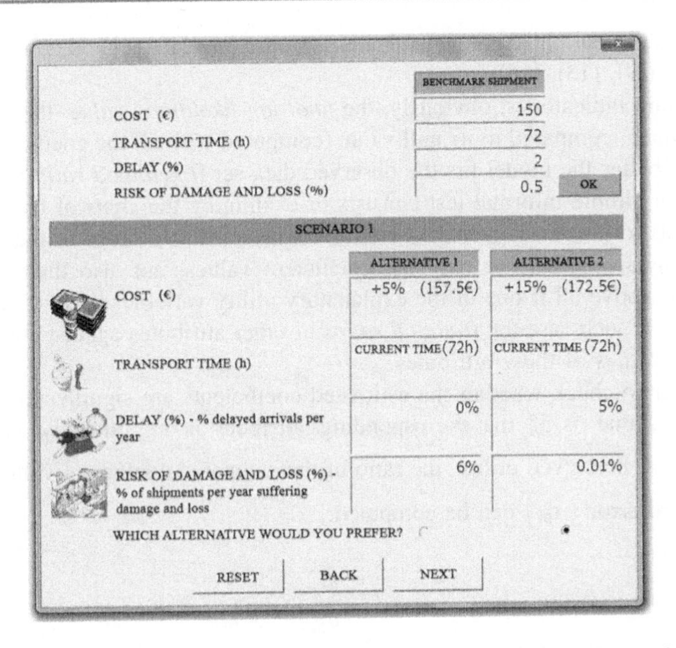


**Fig. 13.5** Input window used to carry out the 20 choice tasks—example of scenario 1. Before starting the choice experiment respondents have to insert their benchmark shipment attribute values

where

- $V_i$ = systematic or representative utility of alternative i.
- $c_i$ = transport cost of alternative i (euro).
- $t_i$ = transport time of alternative i (hours).
- $punct_i$ = punctuality of alternative i (percentage of shipments per year arriving on time), defined as $punct_i = 1 - delay_i$, since the choice tasks were designed with reference to delay levels.
- $risk_i$ = risk of damage and loss for alternative i (percentage of shipments per year suffering damage and/or loss which cause the damaged products to be replaced).
- *ASC* is the *alternative specific constant* representing the net influence of all unobserved, or not explicitly included, characteristics of the options in the utility function. Since the experiments were unlabelled, one would expect the ASC to be zero.

To estimate the coefficients $\beta_k$ associated to each representative attribute the maximum likelihood method was used. The maximum likelihood estimates the set of parameters that have the greatest probability of having generated the sample of independent observations collected with the binary choice experiment.

This estimation has been carried out through the econometric software BIOGEME.

Besides the estimated coefficients, the software provides several goodness-of-fit indicators ([9], [13], [16]).

The main indicator is, obviously, the *final log-likelihood value*: the higher the final estimate, compared to its null value (computed with all the coefficients equal to 0), the better the model fits the observed data set (*likelihood ratio test*).

Another simple informal test consists of examining the *signs* of the estimated coefficients to judge whether they conform with a priori notions or theory.

In many instances, not only the coefficient values, but also their ratios are important, above all if one of the explanatory utility variables is '*cost*', as it is in our study; in such case the *trade-off ratios* of other attributes against cost give the *monetary values* of those attributes.

In order to check whether the estimated coefficients are significantly different from zero, that is if the corresponding attribute is significantly influencing respondents' perceived utility, the ratio of the estimated coefficient value $\left(\hat{\beta}_k\right)$ to its standard error $\left(\sigma_{\beta_k}\right)$ can be computed:

$$t = \frac{\hat{\beta}_k}{\sigma_{\beta_k}}.$$

Sufficiently large values of $t$ (bigger than $\pm 1.96$ for 95 % confidence levels, or than $\pm 1.65$ for 90 % confidence levels) lead to the rejection of the null hypothesis $\hat{\beta}_k = 0$ and hence to accepting that the $k$-th attribute has a significant effect.

Another indicator provided by the estimation software is the $\rho^2$ index:

$$\rho^2 = 1 - \frac{l\left(\hat{\beta}_k, \ldots, \hat{\beta}_k\right)}{l(0)}$$

where

- $l\left(\hat{\beta}_k, \ldots, \hat{\beta}_k\right)$ is the final log-likelihood computed using the final estimated coefficients.
- $l(0)$ is the log-likelihood value when all the model coefficients are equal to zero. This model is known as the equally likely model since it's valid only when none of the chosen attributes can explain the respondents' choices, i.e. when all the proposed alternatives are equiprobable.

  This index, like $R^2$ in regression models, varies between 0 (no model fit since $l\left(\hat{\beta}_k, \ldots, \hat{\beta}_k\right) = l(0)$) and 1 (perfect model fit since the maximum log-likelihood value should tend to zero). However, although its meaning is clear in the limits, it doesn't have an intuitive and unique interpretation for intermediate values; in fact, values around 0.4 may be excellent fits. Furthermore, $\rho^2$ is only appropriate when both alternatives are chosen in the same proportion. To solve this difficulty, a simple adjustment is considered:

$$\bar{\rho}^2 = 1 - \frac{l\left(\hat{\beta}_k, \ldots, \hat{\beta}_k\right)}{l(C)}$$

where

- $l(C)$ is the log-likelihood value when all the model coefficients, except the alternative specific constant, are equal to zero. This model is known as the market share model.

## 13.5 Reference Population and Sample

### 13.5.1 Population

The reference population of our survey was represented by manufacturing firms located in North West Italy: logistics managers in charge of organizing goods shipments (buying ex works or selling cost insurance and freight) for such companies were interviewed. The survey was then targeted at a specific market segment, so its results are not representative for the entire freight market along Corridor Rotterdam-Genoa.

All manufacturing firms were considered excluding those belonging to the first NACE (Rev. 1.1-2002) production sector (DA—manufacture of food products, beverages and tobacco) since the shipments of such perishable products normally show very different characteristics in terms of some of the selected attributes such as load type, acceptable delay, risk of damage, etc. Table 13.2 lists the production sectors considered for the analysis.

Firms with less than 20 employees were also excluded on the hypothesis that they have less important trades than bigger companies.

The distribution by Region of the 14,955 firms included in the reference population just defined is shown in Table 13.3.

It can be noticed that Liguria and Val d'Aosta firms are less than 5 % of the total population so they were further excluded and the sample was restricted only to Piemonte and Lombardia. The new reference population amounted then to 14,469 local units: 74 % are located in Lombardia and the other 26 % in Piemonte.

### 13.5.2 Sample

Conjoint analysis does not necessarily need a random sample: to investigate individual choices for consumer goods or services that are not diffused everywhere it's better to interview people having experienced those products. In particular, freight transport service supply isn't universally available and to get representative observations, interviews with people who already know the analysed services are needed. That is why our sample was restricted to companies buying *ex works* or

**Table 13.2** Production sectors considered in the survey

| Production sector (NACE classification) | |
|---|---|
| DB | Manufacture of textiles and textile products |
| DD | Manufacture of wood and wood products |
| DE | Manufacture of pulp, paper and paper products; publishing and printing |
| DG | Manufacture of chemicals, chemical products and man-made fibres |
| DH | Manufacture of rubber and plastic products |
| DI | Manufacture of other non-metallic mineral products |
| DJ | Manufacture of basic metals and fabricated metal products |
| DK | Manufacture of machinery and equipment n.e.c |
| DL | Manufacture of electrical and optical equipment |
| DN | Manufacturing n.e.c. |

**Table 13.3** Number of local units included in the reference population by region

| Region | TOT LU |
|---|---|
| Piemonte | 3,823 |
| Valle d'Aosta | 34 |
| Lombardia | 10,646 |
| Liguria | 452 |

selling *cost insurance and freight*, since we needed to interview logistics managers in charge of organizing goods shipments for their firms.

With random sampling, sampling error can be reduced by simply increasing the sample size. With nonrandom sampling, however, there is no guarantee that just increasing sample size will make the samples more representative of the population. Furthermore some respondents could resist being interviewed and, by selecting themselves out of the study, they're a source of non-response bias. Measurement error, both with random and nonrandom sampling, can be reduced by having more or better data and observations from each respondent, in particular in conjoint analysis this means including more conjoint questions ([9], [12]).

According to the literature related to conjoint analysis the sample size ($n$ being the number of respondents) should be at least:

$$n > \frac{500 \cdot c}{t \cdot a}$$

where

| | Description | In our case study |
|---|---|---|
| t | Number of tasks | 20 |
| a | Number of alternatives per task | 2 |
| c | Number of parameters to be estimated (coefficients of the selected transport attributes + ASC) | 5 |

Then, in our case: $n > 63$.

In order to select a convenient sample, 358 manufacturing firms were contacted by phone: 251 didn't agree to participate in the survey, 28 were not buying ex works or selling cost insurance and freight firms, 79 agreed to arrange a meeting for the face to face interview. This figure also includes the 4 pilot interviews, carried out before starting the real survey, which were included in the database since they proved to be significant and none of the choice tasks or of the relevant questions in the RP questionnaire were modified after the pilot test.

The localisation of the sample firms is representative of the reference population (Fig. 13.6), Piemonte shows a little higher rate since the 4 pilot interviews included in the final database were all conducted in this region.

On the other hand, the sample is not representative for the chosen market segment as far as the company size (in number of employees—Fig. 13.7) is concerned: more than 50 % of the investigated companies has more than 50 employees and almost 20 % has more than 100 employees while, in the reference population, 65 % of the firms has between 20 and 50 employees and only 14 % has more than 100 employees. The main concern was to include in the survey big companies with a wide range of transport requirements, a specialized logistics department and a high number of suppliers and clients, excluding, at the same time, the firms outsourcing logistics services, for this reason the number of investigated small firms (20–49 employees) was reduced to almost 50 % and the remaining 15 % of interviews was split among the other classes.

Since some of the 79 respondents answered 2 different choice experiments (both for an import and an export shipment), the total number of useful choice experiments included in the final database (neglecting wrong surveys) amounted to 89 which corresponds to 1,651 observations (some of the respondents didn't answer to all the 20 choice tasks of the experiments).

Comparing these figures with the results reported in Table 13.4 obtained from the sample size formula for a finite population:

$$n = \left[\frac{\sigma_p^2 \cdot (N-1)}{N \cdot p \cdot (1-p)} + \frac{1}{N}\right]^{-1}$$

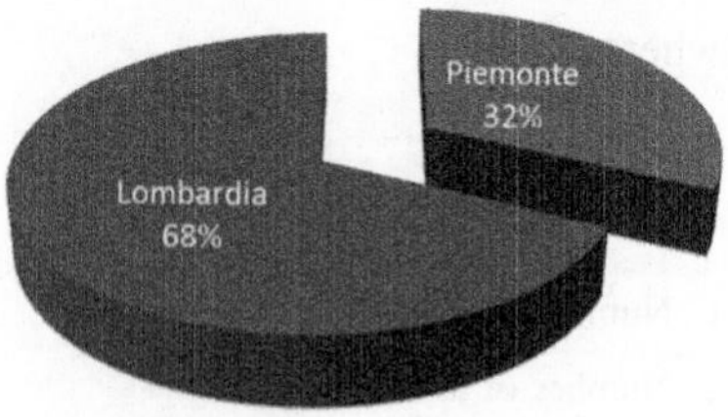

**Fig. 13.6** Percentage distribution of the 79 firms included in the sample by Region

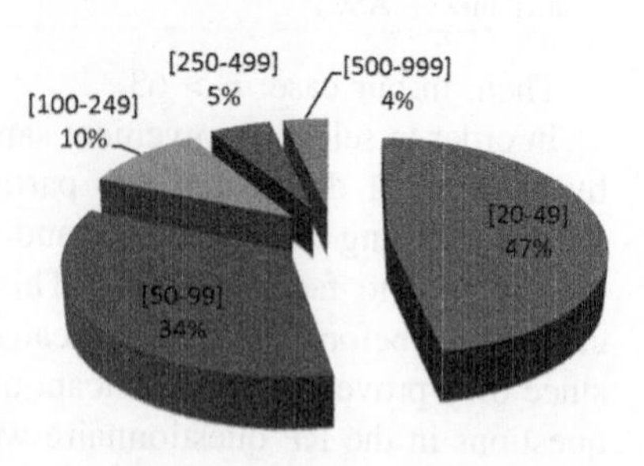


**Fig. 13.7** Percentage distribution of the 79 firms included in the sample by number of employees

it can be stated that the size of our sample is big enough to get significant results from the interviewees' answers: even in the very unlikely case of fully correlated answers (when all choice situations for one respondent collapse to a single observation giving effectively only 89 observations) the sample is still big enough for 90 % confidence level and 10 % confidence interval.

In the formula:

- $N$ is the size of the reference population.
- $\sigma_p^2 = \frac{\Delta^2}{U_c^2}$ is the variance of the sample. $\Delta$ is the confidence interval (precision or sampling error) value, while $U_c$ is the value associated to the percentile of a standard Normal distribution (with mean 0 and typical deviation 1) which entails a probability equal to the confidence level value.
- $p$ is the normally unknown proportion of the attributes being measured in the population; $p = 0.5$ gives the maximum variability level $p(1 - p)$.

## 13.6 Results

### 13.6.1 Revealed Preferences

In this section the results obtained from the elaboration of logistics managers' answers to the RP interviews are described in order to present the main sample attributes and characteristics.

**Table 13.4** Number of respondents needed to satisfy predefined confidence intervals and confidence levels

| | Confidence interval | |
|---|---|---|
| Confidence level (with $p = 0.5$) | 5 % | 10 % |
| 90 % (Uc = 1.65) | 267 | 68 |
| 95 % (Uc = 1.96) | 375 | 95 |
| 99 % (Uc = 2.58) | 637 | 165 |

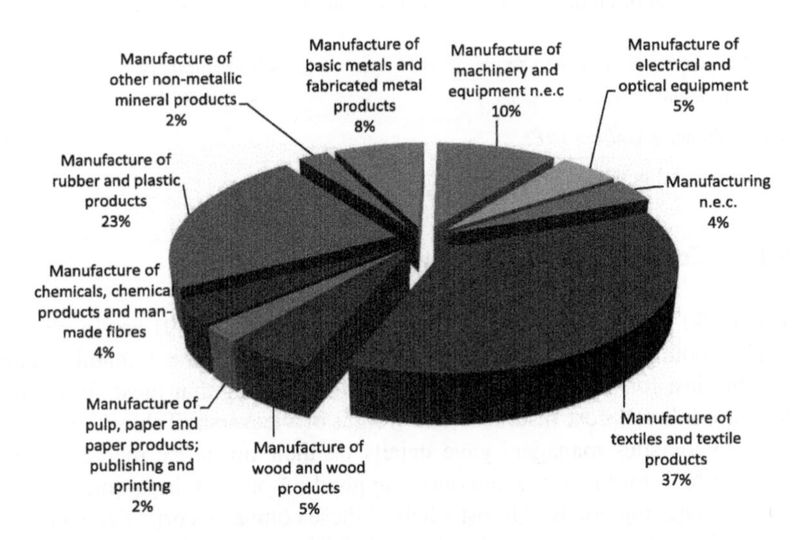


**Fig. 13.8** Production sectors (NACE classification) of the sample firms

#### 13.6.1.1 Company Size and Location

Most of the manufacturing firms that agreed to be interviewed belong to the textile sector (37 % of the sample), followed by the manufacture of rubber and plastic products (23 %), manufacture of machinery (10 %) and manufacture of metal products (8 %) sectors (Fig. 13.8).

Only 32 firms out of 79 declared their average turnover per year and for all of them it's higher than 500,000€. More than a half out of the 32 respondents declared a revenue between 2 and 10 million euros per year, 22 % between 10 and 50 million euros, while the turnover of 16 % of them exceeds 50 million euros per year.

The 79 interviewed firms own a total of 134 production and distribution centres mostly located in Lombardia (49 %) and Piemonte (26 %) where their headquarters are. A significant percentage (16 %) of production plants is located in non-European Countries as Fig. 13.9 shows.

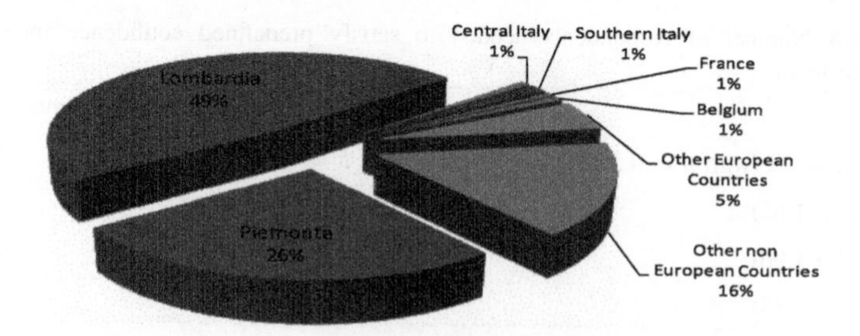


**Fig. 13.9** Location of production and distribution centres of the sample firms

**Table 13.5** Distribution of the sample firms by type of organized shipment

| | | |
|---|---|---|
| Incoming shipments (*buying ex works*) | 6 | 7.6 % |
| Outgoing shipments (*selling CIF*) | 7 | 8.86 % |
| Both incoming and outgoing shipments | 66 | 83.54 % |

#### 13.6.1.2 Company Transport and Logistics

More than 80 % of the interviewed companies (66 out of 79) are in charge of organizing both incoming and outgoing shipments while the remaining part is responsible just for the import or for the export, meaning that these firms buy ex works but don't sell cost insurance and freight or viceversa (Table 13.5).

Only 73 logistics managers gave details on their inventory management; 22 described it both for incoming and outgoing goods, 5 only for incoming goods and 46 only for outgoing goods. Almost a half of these companies organize their stocks by replenishing them on received orders (47 % of the sample), 23 % maintain the warehouse goods level greater than or equal to a threshold reorder point, the remaining quotas are split between firms replenishing their stocks on production scheduling and firms using just in time production strategies.

Only some of the respondents provided details about the modal split of their company's shipments that was calculated on the basis of:

- The overall tonnes transported per year (17 valid answers for incoming goods and 30 for outgoing goods).
- The total number of shipments per year (21 valid answers for incoming shipments and 25 for outgoing shipments).

The collected values are reported in Table 13.6, while the corresponding percentages are shown in Figs. 13.10 and 13.11.

Even though the data do not refer to the overall sample, but only to shipments or to transported tonnes (not always matched together) declared by a portion of the respondents, the modal split distribution is quite similar to the average Italian trend which assigns to road transport services the most of shipped goods. According to the received answers, more than 82 % of the yearly shipments is carried out by

**Table 13.6** Modal split for revealed tonnes/year and revealed transports/year

| | Revealed tonnes/year | Number of revealed transports/year |
|---|---|---|
| Road | 432,338.7 | 54,984.5 |
| Rail | 3.2 | – |
| Road + rail | 23,140.0 | 1,119.0 |
| Road + inland waterways | – | – |
| Road + air | 1,046.4 | 6,820.0 |
| Road + sea | 29,895.7 | 3,486.5 |
| Rail + sea | – | – |
| Other | – | – |
| TOT | 486,424.0 | 66,410.0 |

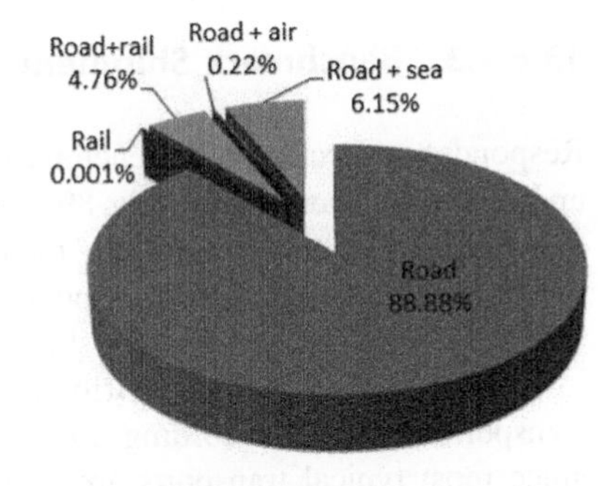


**Fig. 13.10** Modal split calculated on the revealed transported tonnes/year

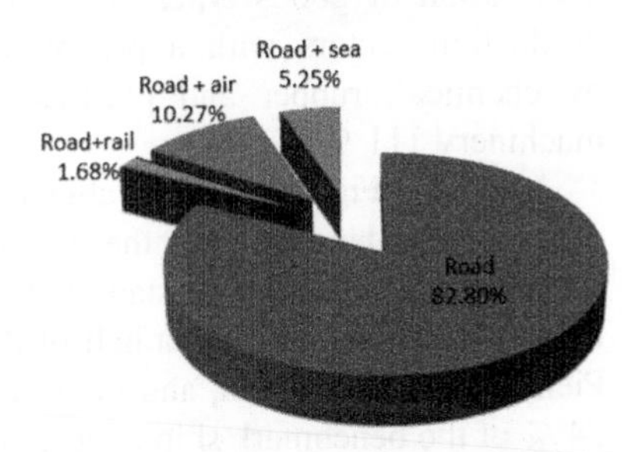


**Fig. 13.11** Modal split calculated on the revealed total shipments/year

road, while the remaining shares use intermodal transport services, in particular road + air (10 %), road + sea (5 %) and road + rail (2 %) services. Looking at goods volumes (tonnes/year), an higher percentage of goods transported by road (89 %) can be noticed, on the other hand the percentage value of tonnes transported by road + air mode decreases significantly (0.22 %), since goods volumes shipped by plane are generally quite small.

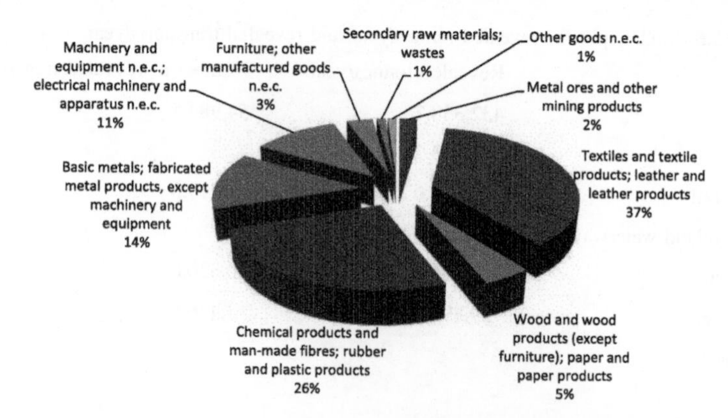


**Fig. 13.12** Distribution of the benchmark shipments by type of transported goods—NST2007 classification

#### 13.6.1.3 Benchmark Shipment

Respondents described 97 benchmark shipments (18 of them both for incoming and outgoing goods), but only 89 out of 97 were related to a significant choice experiment, for this reason the main attributes of benchmark shipments were analysed with reference only to the 89 valid answers: 18 % referred to an incoming shipment, 82 % to an outgoing shipment.

Figure 13.12 shows the distribution of the 89 benchmark shipments by type of transported goods, according to NST2007 goods classification. As expected, since most typical transports are referred to outgoing shipments, the percentage distribution of goods types is almost the same as the distribution of the firms production sectors with a predominance of textile products (37 %), followed by chemical, rubber and plastics products (26 %), basic metals (14 %) and machinery (11 %).

Analysing Fig. 13.13 and Table 13.7 which show the distribution of origins and destinations (depending on the type of transport described) of the 89 benchmark shipments, it's possible to state that more than 60 % of the benchmark shipments is carried out within Italy: a half of them (31 %) remains within the study area, in Piemonte or Lombardia, and the others go to or come from other Italian regions. 24 % of the benchmark shipments are directed to other European Countries among whom Germany and France are the most popular (10 % of the shipments in German direction and 3 % to or from France); the remaining 14 % of typical transports are going to or coming from other non-European Countries.

Modal split of the overall benchmark shipments, shown in Fig. 13.14, looks very similar to the transport mode distribution (calculated on the total number of firm shipments per year) stated by the respondents in the questionnaire section about transport and logistics. 82 % of typical transports (73 out of 89) is carried out by road, while the remaining quota by combined transport; in this case also

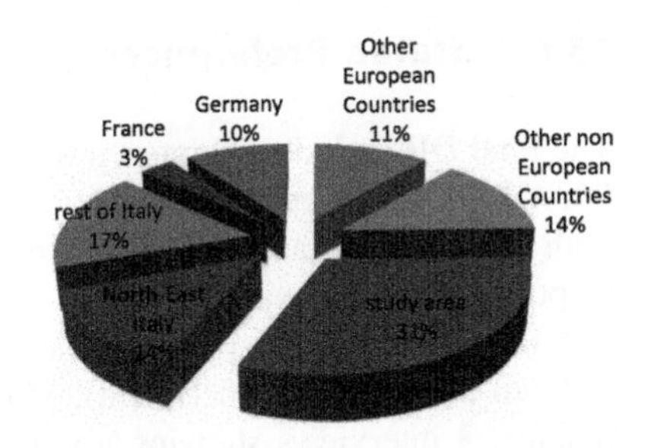


**Fig. 13.13** Origin or destination of the benchmark shipments

**Table 13.7** Number of benchmark shipments by origin or destination area

| Zone of origin or destination (depending on the described benchmark shipment) | Number of shipments |
|---|---|
| Study area | 28 |
| North-East Italy | 12 |
| Rest of Italy | 15 |
| France | 3 |
| Germany | 9 |
| Other European countries | 10 |
| Other non European countries | 12 |

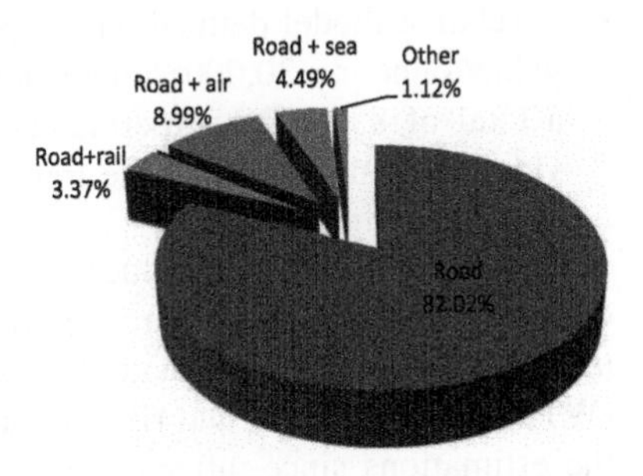


**Fig. 13.14** Transport mode of the benchmark shipments

other modes are included since one of the described benchmark shipments uses more than two transport modes (road + sea + rail). 79 % of the benchmark shipments are general cargo shipments, 10 % move unitized goods like containers or swap bodies and 8 % move liquid or solid bulks. The remaining 3 % (3 shipments) declared other load types (e.g. marbles, concrete or machinery).

90 % of the shipments are carried out by third parties and only 10 % (9 shipments) are own account shipments.

### 13.6.2 Stated Preferences

The final DB includes 79 interviews to manufacturing firms located in North West Italy: 18 of the interviewed logistics managers described 2 benchmark shipments, both for incoming and outgoing goods, but only 13 of them completed the 2 corresponding choice experiments, so we collected (79 + 13) 92 choice experiments and 1,697 observations (some of the respondents didn't answer to all the 20 choice tasks of each choice experiment). Moreover, in order to estimate the model coefficients, 3 interviews showing a wrong choice for the dominant alternative were removed; this led to analyse 89 choice experiments and a total of 1,651 observations.

Five different statistical model estimations were carried out through:

- A relative model using the attribute levels indicated in Table 13.1 (relative values for transport cost and time), based on the 89 valid choice experiments.
- An absolute model using absolute attribute levels (real transport costs and times computed with reference to the described benchmark shipment values) based on the 89 valid choice experiments.
- A relative model defined with separate coefficients by road and intermodal mode (using dummy variables and keeping the cost coefficient constant) based on the 89 valid choice experiments.
- A relative model defined with separate coefficients by road and intermodal mode and by consolidated shipments (load weight <= 5 tonnes) and truckload (load weight >5 tonnes) based on the 89 valid choice experiments.
- A relative model defined with separate coefficients by goods value/tonne (low value/tonne <= 30,000€/tonne and high value/tonne >30,000€/tonne) based on a total of 85 choice experiments (referred to benchmark shipments that provided information about the transported goods value) and 1,574 observations.

The average attribute values per shipment derived from the different subsets considered in the models are reported in Fig. 13.15, the highlighted values are used as a reference to evaluate the value of time (VoT) and the willingness to pay (WTP) for punctuality and risk of damage and loss. No outliers were removed for the estimations since, after double checking collected data, all the observations proved to be reliable or explicable.

Average values are quite reasonable: compared to intermodal trips, road shipments show lower transport cost and time as well as a smaller number of shipments per year (in %) arriving late or suffering damage and loss. According to the average transport length, also the acceptable delay (in hours) for road transport mode is lower than for intermodal mode.

Regarding transport cost, as expected, intermodal shipments have lower cost/km than road trips, while they show a high average cost/tonne value, which could be explained by the fact that several road + air shipments are included in the database. The average cost/km and cost/tonne values per shipment, both for road and intermodal mode, roughly agree (excluding some exceptions) with the values obtained by a benchmark survey on freight transport carried out by the Bank of

| | AVERAGE VALUES PER SHIPMENT | | | | | | |
|---|---|---|---|---|---|---|---|
| | ALL BENCHMARK SHIPMENTS (89) | ONLY ROAD SHIPMENTS (73) | ONLY ROAD SHIPMENTS WITH LIGHT LOAD (< 5tonnes) | ONLY ROAD SHIPMENTS WITH HEAVY LOAD (> 5tonnes) | ONLY INTERMODAL SHIPMENTS (16) | LOW VALUE/TONNE SHIPMENTS (<30000 €/tonne) | HIGH VALUE/TONNE SHIPMENTS (>30000 €/tonne) |
| **Transport cost (€)** | 622.38 | 361.53 | 191.85 | 811.20 | 1812.50 | 801.44 | 293.77 |
| **Transport time (h)** | 91.70 | 40.68 | 47.21 | 23.38 | 324.50 | 85.63 | 107.60 |
| **Acceptable delay (%)** - max % of additional transport time that you would accept in order to consider the shipment still on time | 23.78% | 25.55% | 26.33% | 21.39% | 22.76% | 27.01% | 19.30% |
| **Acceptable delay (h)** | 21.80 | 10.39 | 12.43 | 5.00 | 73.86 | 23.13 | 20.76 |
| **Delay (%)** - % of delayed arrivals per year | 5.26% | 4.38% | 6.36% | 1.15% | 12.38% | 2.34% | 8.41% |
| **Delay (number of shipments)** - number of delayed arrivals per year | 18.69 | 16.84 | 20.87 | 6.15 | 27.13 | 6.98 | 41.30 |
| **Risk of damage and loss (%)** - % of shipments per year suffering damage and loss which cause the damaged products to be replaced | 0.52% | 0.48% | 0.71% | 0.11% | 0.81% | 0.25% | 0.81% |
| **Risk of damage and loss (number of shipments)** - number of shipments per year suffering damage and loss which cause the damaged products to be replaced | 1.83 | 1.85 | 2.33 | 0.58 | 1.77 | 0.73 | 3.98 |
| **Number of shipments per year** | 355 | 385 | 328 | 535 | 219 | 298 | 491 |
| **Distance (km)** | 1678.49 | 482.41 | 483.72 | 478.95 | 7135.63 | 1563.41 | 2032.10 |
| **Weight of goods (tonnes)** | 6.93 | 7.02 | 0.77 | 24.45 | 6.52 | 10.70 | 0.38 |
| **Tonnes*km** | 8524.84 | 3084.95 | 307.34 | 10833.00 | 33004.39 | 12542.97 | 1647.64 |
| **Transport cost/h (€/h)** | 6.79 | 8.89 | 4.06 | 34.70 | 5.59 | 9.36 | 2.73 |
| **Transport cost/km (€/km)** | 0.37 | 0.75 | 0.40 | 1.69 | 0.25 | 0.51 | 0.14 |
| **Transport cost /tonnes (€/tonne)** | 88.38 | 49.26 | 248.84 | 21.89 | 277.83 | 74.93 | 766.20 |
| **Transport cost/tonnes*km** | 0.07 | 0.11 | 0.62 | 0.05 | 0.05 | 0.06 | 0.18 |
| **Speed (km/h)** | 18.30 | 11.86 | 10.25 | 20.49 | 21.99 | 18.26 | 18.89 |
| **Value of goods (€)** | 28507.53 | 22737.71 | 16898.82 | 38410.53 | 55433.33 | 29483.15 | 26808.06 |
| **Value of goods per tonne (€/tonne)** | 4110.92 | 3151.39 | 21249.04 | 1571.15 | 9854.11 | 2756.61 | 69918.39 |

**Fig. 13.15** Average shipment characteristics. The *highlighted cells* are the model reference values used to evaluate VoT and WTP

Italy in 2008. The average cost/tonnes km values are comparable to the results of a survey similar to ours conducted in three Italian regions in 2006 by Danielis and Marcucci [3]. That paper reports results, by production sector and firm size, that vary from 0.01€/tonne km (metallurgic factories with 51–100 employees) to 0.5€/tonne km (furniture factories with less than 50 employees). Not all the production sectors are represented.

What may appear curious is the average speed value, calculated as the weighted average of the velocities, $v = \sum_i \text{distance}_i / \sum_i \text{time}_i$, since intermodal shipments seem to be faster on average than road ones. This could be explained partly by the fact that intermodal shipments include, besides road + rail and road + sea transports, examples of faster shipments carried out by road + air means, partly by the fact that the average road shipments speed is very low (12 km/h), too much to be explained by resting times rules, by loading and unloading times or by waiting times at borders, tunnels, etc.

Analyzing these road transport data, it has been observed that shipments lasting less than one day show a quite acceptable average speed (33 km/h), while speed quickly decreases for more time-consuming trips. The average speed values per road shipment by transport time are shown in Table 13.8.

**Table 13.8** Average speed per road shipment by transport time

| Number of shipments | Transport time per shipment (h) | Average speed (km/h) |
|---|---|---|
| 28 | <24 | 33.26 |
| 15 | 24 | 13.06 |
| 1 | 36 | 24.72 |
| 15 | 48 | 10.81 |
| 6 | 72 | 11.68 |
| 4 | 96 | 14.43 |
| 2 | 120 | 10.21 |
| 2 | >120 | 6.89 |

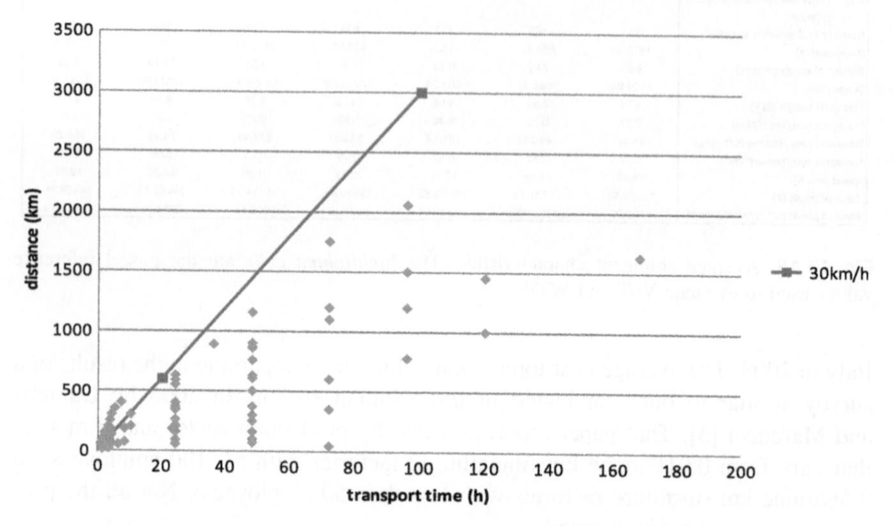


**Fig. 13.16** Distance-transport-time graph for road transport benchmark shipments

It's worth noticing also that, when transport time exceeds 24 h, respondents usually indicated it in days and not in hours: this led to have numerous shipments taking 24/48/72 or 96 h to cover very different distances; for instance: the 24 h shipments in our database cover from 55 to 630 km (Fig. 13.16).

Most of the very short but time-consuming road shipments included in the database (which cause the speed of this category to decrease) carry low weight goods (<1 tonne). These shipments could then be carried out through groupage (consolidated) services which take more delivery time than direct truck shipments and normally carry low weight packages thus having high costs/tonne while decreasing costs/h and costs/km (e.g.: one of the interviewed shippers, importing rubber materials described a benchmark shipment costing 30€ and taking 48 h to cover 70 km; it seems quite strange but, since the transported goods weight only 20 kg, this

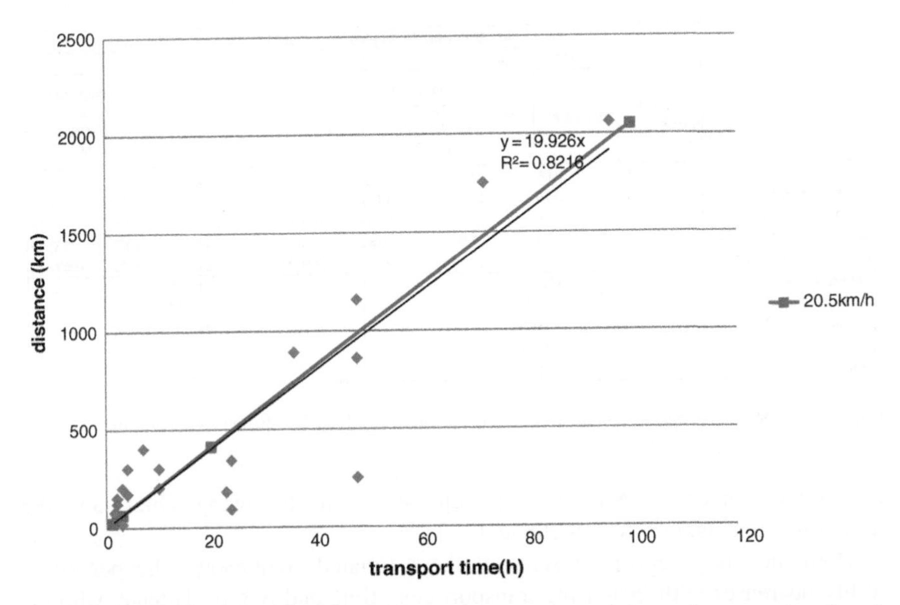


**Fig. 13.17** Distance-transport-time and average speed value graph for road transport benchmark shipments weighting more than 5 tonnes (truck shipments)

could be a consolidated shipment. The shipper, who was asked for transport time and covered distance, only knows that his goods arrive at his warehouse 2 days after their departure from the origin, without knowing the real tour made by the carrier who probably covered a longer path delivering also other goods).

Though the average goods weight in our database is higher than 6 tonnes, there're 51 benchmark shipments (out of 89) weighting less than 1 tonne and 57 weighting less than 3 tonnes; the goods weight median value is in fact 0.5 tonnes. If we analyze only road transport shipments weighting more than 5 tonnes (Fig. 13.17), that is if we exclude all possible groupage shipments, the average speed is then 20.5 km/h, a more reliable value.

Furthermore other studies on the same topic found similar average speed values: Zotti and Danielis [17] studied the mechanics manufactures in Friuli Venezia Giulia, Maggi and Rudel [8] interviewed 35 shippers of medium and large companies of food and wholesale sector in Switzerland and the average benchmark shipment speed they found was 14.8 km/h (Zotti—only road mode) and 14.5 km/h (Maggi—all modes).

#### 13.6.2.1 Estimations on the Overall 89 Benchmark Shipments with Both Relative and Absolute Models

Figure 13.18 shows the final results obtained through both a relative and an absolute estimation model and includes estimated coefficients, standard errors, *t* test values (robust *t* tests and standard errors, which partially correct the problem

| | ALL BENCHMARK SHIPMENTS - 89 INTERVIEWS | | | | | | | |
|---|---|---|---|---|---|---|---|---|
| | RELATIVE MODEL | | | | ABSOLUTE MODEL | | | |
| | COEFFICIENT VALUE | STD ERROR | T-TEST | TRADE OFF RATIO AGAINST COST | COEFFICIENT VALUE | STD ERROR | T-TEST | TRADE OFF RATIO AGAINST COST |
| ASC | 0.25 | 0.06 | 3.88 | | 0.1990 | 0.05 | 3.62 | |
| βc - TRANSPORT COST | -13.00 | 0.76 | -17.04 | 1.00 | -0.0028 | 0.00 | -2.49 | 1.00 |
| βt - TRANSPORT TIME | -2.67 | 0.36 | -7.47 | 0.21 | -0.0019 | 0.00 | -1.63 | 0.68 |
| βpunct - PUNCTUALITY | 5.76 | 0.64 | 8.98 | -0.44 | 2.2800 | 0.49 | 4.67 | -802.82 |
| βrisk - RISK OF DAMAGE | -46.50 | 2.69 | -17.27 | 3.58 | -29.8000 | 1.95 | -15.30 | 10492.96 |
| Null log-likelihood | | -1144.39 | | | | -1144.39 | | |
| Cte log-likelihood | | -1130.87 | | | | -1130.87 | | |
| Final log-likelihood | | -756.98 | | | | -961.91 | | |
| ρ2 | | 0.34 | | | | 0.16 | | |
| Adjusted ρ2 | | 0.33 | | | | 0.16 | | |

**Fig. 13.18** Relative and absolute model estimations based on 89 choice experiments

of not independent observations, are reported in the following tables) and the trade-off ratios against cost coefficient.

Both models give correct signs of the estimated coefficients: the perceived utility decreases with increasing transport cost, time and risk of damage, while it increases if the punctuality increases.

The final log-likelihood and the rho-square values are rather satisfactory and show a significant though not very large improvement in the relative model, while a quite low improvement is shown by the absolute model.

The values of the *t* test in the relative model show that all estimates are significant while in the absolute models the trip time attribute is not even significant at the 10 % level. The ASC had better have a non-significant *t* test value since the alternatives are unlabelled and respondents shouldn't have an a priori preference between them.

Looking at the trade-off ratios against the cost coefficient provided by the relative model, it's possible to state that a 10 % increase in transport time is equivalent to a 2 % higher transport cost ($\beta_t/\beta_c = 0.21$), while a 10 % increase in punctuality (e.g. from 90 to 99 % of shipments/year arriving on time) is equivalent to a 4 % decrease in transport cost and a 10 % increase in risk of damage and loss (e.g. from 2 to 2.2 %) is equivalent to a 36 % higher transport cost. These results also mean that increasing punctuality counts almost 2 times as much as decreasing transport time by the same percentage, while increasing risk of damage counts 17 times as much as increasing transport time.

Observing the absolute model trade-off ratios and statistical indicators, it's evident that the relative model proved to be clearly better than the absolute one so all the following considerations and analysis were carried out referring to the relative model.

With reference to the estimated coefficients and trade-off ratios, the monetary values of the three quality attributes used in the SP experiments were computed (also using the attribute values from Fig. 13.15); they correspond to the willingness to pay (WTP) for an improved quality of the average shipment attributes or

| ALL BENCHMARK SHIPMENTS - 89 INTERVIEWS | | |
|---|---|---|
| WILLINGNESS TO PAY | RELATIVE MODEL | |
| | €/transport | €/tonne |
| Reduction of 10% in transport time | 12.78 | 1.84 |
| Reduction of 1h in transport time | 1.39 | 0.20 |
| Reduction of 1day in transport time | 33.46 | 4.83 |
| Increase of 10% in the number of shipments/year arriving on time | 27.60 | 3.98 |
| Increase of 5 shipments/year arriving on time | 3.88 | 0.56 |
| Reduction of 10% in the number of shipments/year suffering damage and/or loss | 222.60 | 32.12 |
| Reduction of 5 shipments/year suffering damage and/or loss | 31.36 | 4.52 |

**Fig. 13.19** WTP for an improved quality of the average shipment attributes computed on the basis of 89 choice experiments

similarly, given the model symmetry, to the willingness to accept (WTA) a decreased quality of the same attributes. Results are shown in Fig. 13.19.

The value of time (VoT) is lower than the monetary values of the other quality attributes meaning that transport duration is of limited importance for shippers organizing mostly third parties account shipments. This happens because shippers' VoT is only related to the cargo, not to the staff and vehicle costs; a shorter transport time only decreases inventory costs during the shipment, while a higher number of shipments arriving late or suffering damages produces bigger losses (interruption of the production chain/loss of reliability as a supplier, etc.).

An attempt to estimate two different coefficients per attribute related to attribute gains or losses (i.e. whether attributes increase (inc) or decrease (dec) compared to their reference levels) was made. The systematic utility function was defined as follows, keeping the last two attributes (already defined without a reference value) symmetric:

$$U_i = ASC_i + \sum_{k=c,t} \beta_k(dec)x_{ik}(dec) + \sum_{k=c,t} \beta_k(inc)x_{ik}(inc) + \beta_p punct_i + \beta_r risk_i$$

where $x_{ik} = c_i, t_i$ for $k = c,\ t$ and $x_{ik}(dec) = \max(x_{ref} - x_{ik}, 0)$ and $x_{ik}(inc) = \max(x_{ik} - x_{ref}, 0)$.

The trade-off ratios were evaluated as follows:

- For asymmetric undesirable attributes (e.g. transport time): $WTP \Rightarrow \frac{\beta_k(dec)}{\beta_c(inc)}$; $WTA \Rightarrow \frac{\beta_k(inc)}{\beta_c(dec)}, \quad k = t.$
- For symmetric attributes (e.g. punctuality and risk of damage): $\frac{\beta_k}{\beta_c}, \quad k = p, r.$

The estimation of such model provided some wrong results since not all the estimated coefficient signs proved to be correct. In particular, both time coefficients (for losses and gains) were negative, while the coefficients of all other attributes showed a correct sign: positive for decreasing values and negative for increasing values for cost, negative for risk of damage and positive for punctuality.

| | ALL BENCHMARK SHIPMENTS - 89 INTERVIEWS | | | |
|---|---|---|---|---|
| | COEFFICIENT VALUE | STD ERROR | T-TEST | TRADE OFF RATIO AGAINST COST |
| ASC | 0.253 | 0.065 | 3.890 | |
| βc - TRANSPORT COST | -13.000 | 0.766 | -16.980 | 1.00 |
| βt_int - INTERMODAL TRANSPORT TIME | -3.100 | 0.787 | -3.940 | 0.24 |
| βt_road - ROAD TRANSPORT TIME | -2.610 | 0.392 | -6.640 | 0.20 |
| βpunct_int - INTERMODAL PUNCTUALITY | 4.480 | 1.460 | 3.060 | -0.34 |
| βpunct_road - ROAD PUNCTUALITY | 6.090 | 0.704 | 8.660 | -0.47 |
| βrisk_int - INTERMODAL RISK OF DAMAGE | -62.800 | 7.490 | -8.390 | 4.83 |
| βrisk_road - ROAD RISK OF DAMAGE | -43.800 | 2.800 | -15.660 | 3.37 |
| Null loglikelihood | -1144.386 | | | |
| Cte loglikelihood | -1130.866 | | | |
| Final loglikelihood | -749.483 | | | |
| ρ2 | 0.345 | | | |
| Adjusted ρ2 | 0.338 | | | |

**Fig. 13.20** Relative model estimations, with separate coefficients by mode, based on 89 choice experiments

Otherwise the $t$ test evaluated for the wrong transport time coefficient ($\beta_t(dec)$) was equal to −2.27, a small value near the thresholds of statistical significance: probably the sample size is too small and doesn't include enough observations to get correct estimations for all these parameters.

#### 13.6.2.2 Estimations on the Overall 89 Benchmark Shipments with Separate Coefficients by Road and Intermodal Mode

Another model was estimated to take into account different transport modes: separate coefficients by road and intermodal mode were estimated for all quality attributes, keeping the same coefficient only for cost (on the hypothesis that shippers using road or intermodal services have the same perception of transport cost):

$$U_i = ASC_i + \beta_c c_i + \sum_{k=t,p,r} \beta_{k\,road} x_{ik} d_{road} + \sum_{k=t,p,r} \beta_{k\,in} x_{ik} d_{in}$$

where $x_{ik} = t_i, punct_i, risk_i$ for $k = t,\ p,\ r$

$$d_{road} = \begin{cases} 1 & \text{if road mode is chosen in the reference shipment,} \\ 0 & \text{otherwise} \end{cases}$$

$$d_{in} = \begin{cases} 1 & \text{if intermodal mode is chosen in the reference shipment,} \\ 0 & \text{otherwise} \end{cases}$$

Results are shown in Fig. 13.20.

| ALL BENCHMARK SHIPMENTS - 89 INTERVIEWS | | |
|---|---|---|
| WILLINGNESS TO PAY | €/transport | €/tonne |
| Reduction of 10% in ROAD transport time | 7.258 | 1.034 |
| Reduction of 1h in ROAD transport time | 1.784 | 0.254 |
| Reduction of 1day in ROAD transport time | 42.822 | 6.100 |
| Reduction of 10% in INTERMODAL transport time | 43.221 | 6.629 |
| Reduction of 1h in INTERMODAL transport time | 1.332 | 0.204 |
| Reduction of 1day in INTERMODAL transport time | 31.966 | 4.903 |
| Increase of 10% in the number of ROAD shipments/year arriving on time | 16.936 | 2.413 |
| Increase of 5 ROAD shipments/year arriving on time | 2.200 | 0.313 |
| Increase of 10% in the number of INTERMODAL shipments/year arriving on time | 62.462 | 9.580 |
| Increase of 5 INTERMODAL shipments/year arriving on time | 14.261 | 2.187 |
| Reduction of 10% in the number of ROAD shipments/year suffering damage and/or loss | 121.808 | 17.352 |
| Reduction of 5 ROAD shipments/year suffering damage and/or loss | 15.819 | 2.253 |
| Reduction of 10% in the number of INTERMODAL shipments/year suffering damage and/or loss | 875.577 | 134.291 |
| Reduction of 5 INTREMODAL shipments/year suffering damage and/or loss | 199.903 | 30.660 |

**Fig. 13.21** WTP for an improved quality of the average shipment attributes computed on the basis of 89 choice experiments

Estimated coefficients show correct signs. The final log-likelihood and the rho-square values are satisfactory and show a significant improvement. Observing the *t* test values, all estimates prove to be significant.

The trade-off ratios are not much dissimilar to those of the former model. On the basis of these ratios and with reference to the characteristics of the average road and intermodal shipments (reported in Fig. 13.15), shippers' WTP both per transport and per tonne were computed (Fig. 13.21). Shippers using intermodal services would pay more than shippers transporting by road to reduce or increase the same quality attributes by the same percentage.

It may appear strange that transport duration is more important for intermodal than for road shippers (intermodal VoT, referred to the same percentage of time reduction, is higher than road one) but looking at the values reported in Fig. 13.15, it can be seen that intermodal shipments are on average faster than road shipments: this is explained because in our database road transports include many consolidated shipments, moreover intermodal shipments include also road + air mode which considerably increases the average speed. On the other hand, with reference to 1 h or 1 day of transport time, road shippers' willingness to pay is higher than intermodal shippers' one since intermodal transports are definitely longer.

These VoT results can be compared with some results reported by de Jong et al. [5], de Jong [6] and Significance et al. [15]. This shows that the results found in North West Italy for intermodal transport modes are reasonably in line with valuations of shippers for the time related to the goods in transit in Sweden, Norway, The Netherlands and the UK, whereas those for road transport are relatively low.

| | ALL BENCHMARK SHIPMENTS - 89 INTERVIEWS | | | |
|---|---|---|---|---|
| | COEFFICIENT VALUE | STD ERROR | T-TEST | TRADE OFF RATIO AGAINST COST |
| ASC | 0.253 | 0.0653 | 3.88 | |
| βc - TRANSPORT COST | -13.100 | 0.771 | -16.94 | 1.00 |
| βt_int - INTERMODAL TRANSPORT TIME | -3.110 | 0.788 | -3.95 | 0.24 |
| βt_road heavy - ROAD TRANSPORT TIME for heavy loads | -2.090 | 0.676 | -3.09 | 0.16 |
| βt_road light - ROAD TRANSPORT TIME for light loads | -2.820 | 0.463 | -6.08 | 0.22 |
| βpunct_int - INTERMODAL PUNCTUALITY | 4.490 | 1.46 | 3.07 | -0.34 |
| βpunct_road heavy - ROAD PUNCTUALITY for heavy loads | 5.310 | 1.23 | 4.34 | -0.41 |
| βpunct_road light - ROAD PUNCTUALITY for light loads | 6.430 | 0.832 | 7.73 | -0.49 |
| βrisk_int - INTERMODAL RISK OF DAMAGE | -62.900 | 7.5 | -8.4 | 4.80 |
| βrisk_road heavy - ROAD RISK OF DAMAGE for heavy loads | -32.700 | 4.33 | -7.55 | 2.50 |
| βrisk_road light - ROAD RISK OF DAMAGE for light loads | -48.700 | 3.44 | -14.15 | 3.72 |
| Null loglikelihood | -1144.386 | | | |
| Cte loglikelihood | -1130.866 | | | |
| Final loglikelihood | -744.615 | | | |
| $\rho^2$ | 0.349 | | | |
| Adjusted $\rho^2$ | 0.34 | | | |

**Fig. 13.22** Relative model estimations, with separate coefficients by mode and type of road shipment, based on 89 choice experiments

### 13.6.2.3 Estimations on the Overall 89 Benchmark Shipments with Separate Coefficients by Road and Intermodal Mode and by Consolidated Shipments (Load Weight <= 5 tonnes) and Truckload (Load Weight >5 tonnes)

In order to analyse more in depth the monetary values obtained with the previous model a new estimation was carried out considering separate coefficients not only by transport modes but also by consolidated shipments (transporting goods weighting less than 5 tonnes—light loads) and full truck load haulages (transporting goods weighting more than 5 tonnes—heavy loads):

$$U_i = ASC_i + \beta_c c_i + \sum_{k=t,p,r} \beta_{k\,road\,light} x_{ik} d_{road\,light} + \sum_{k=t,p,r} \beta_{k\,road\,heavy} x_{ik} d_{road\,heavy} + \sum_{k=t,p,r} \beta_{k\,in} x_{ik} d_{in}$$

where

$$d_{road\,light} = \begin{cases} 1 & \text{if road mode is chosen in the reference shipment and cargo weights } \leq 5\text{ tonnes,} \\ 0 & \text{otherwise} \end{cases}$$

$$d_{road\,heavy} = \begin{cases} 1 & \text{if road mode is chosen in the reference shipment and cargo weights } > 5\text{ tonnes,} \\ 0 & \text{otherwise} \end{cases}$$

Estimation results and WTP values are shown in Figs. 13.22 and 13.23.

| ALL BENCHMARK SHIPMENTS - 89 INTERVIEWS | | |
|---|---|---|
| **WILLINGNESS TO PAY** | **€/transport** | **€/tonne** |
| **Reduction of 10% in ROAD transport time - light loads (consolidated shipments)** | 4.130 | 5.357 |
| **Reduction of 1h in ROAD transport time - light loads (consolidated shipments)** | 0.875 | 1.135 |
| **Reduction of 1day in ROAD transport time - light loads (consolidated shipments)** | 20.995 | 27.231 |
| **Reduction of 10% in ROAD transport time - heavy loads (truckload)** | 12.942 | 0.529 |
| **Reduction of 1h in ROAD transport time - heavy loads (truckload)** | 5.537 | 0.226 |
| **Reduction of 1day in ROAD transport time - heavy loads (truckload)** | 132.881 | 5.435 |
| **Reduction of 10% in INTERMODAL transport time** | 43.030 | 6.596 |
| **Reduction of 1h in INTERMODAL transport time** | 1.326 | 0.203 |
| **Reduction of 1day in INTERMODAL transport time** | 31.825 | 4.878 |
| **Increase of 10% in the number of ROAD shipments/year arriving on time - light loads (consolidated shipments)** | 9.417 | 12.214 |
| **Increase of 5 ROAD shipments/year arriving on time - light loads (consolidated shipments)** | 1.435 | 1.862 |
| **Increase of 10% in the number of ROAD shipments/year arriving on time - heavy loads (truckload)** | 32.881 | 1.345 |
| **Increase of 5 ROAD shipments/year arriving on time - heavy loads (truckload)** | 3.073 | 0.126 |
| **Increase of 10% in the number of INTERMODAL shipments/year arriving on time** | 62.123 | 9.523 |
| **Increase of 5 INTERMODAL shipments/year arriving on time** | 14.183 | 2.174 |
| **Reduction of 10% in the number of ROAD shipments/year suffering damage and/or loss - light loads (consolidated shipments)** | 71.321 | 92.507 |
| **Reduction of 5 ROAD shipments/year suffering damage and/or loss - light loads (consolidated shipments)** | 10.871 | 14.101 |
| **Reduction of 10% in the number of ROAD shipments/year suffering damage and/or loss - heavy loads (truckload)** | 202.490 | 8.283 |
| **Reduction of 5 ROAD shipments/year suffering damage and/or loss - heavy loads (truckload)** | 18.924 | 0.774 |
| **Reduction of 10% in the number of INTERMODAL shipments/year suffering damage and/or loss** | 870.277 | 133.400 |
| **Reduction of 5 INTREMODAL shipments/year suffering damage and/or loss** | 198.693 | 30.457 |

**Fig. 13.23** WTP for an improved quality of the average shipment attributes computed on the basis of 89 choice experiments

These last estimations show correct signs and the best rho-square value. The *t* test values confirm the statistical significance of estimated coefficients.

Looking at the WTP values it can be noticed that, as expected, the VoT per transport for truckload shipments (faster than consolidated shipments but still a little bit slower than intermodal ones) is higher than that of groupage transports. This is valid also for the WTP for punctuality and risk of damage and loss. These results hold since they're obtained with reference to the different average reference values (for groupage, truckload and intermodal shipments) which show different transport costs. If we look at the trade-off ratios, we find out that the relative importance of transport attributes for heavy loads is a little bit lower than for light loads: this may have to do with the type of transported goods. Truck load cargoes could be more bulky products of lower value (see Fig. 13.15), so, there would be

| | ALL BENCHMARK SHIPMENTS reporting the goods value - 85 INTERVIEWS | | | |
|---|---|---|---|---|
| | COEFFICIENT VALUE | STD ERROR | T-TEST | TRADE OFF RATIO AGAINST COST |
| ASC | 0.254 | 0.066 | 3.840 | |
| βc - TRANSPORT COST | -13.200 | 0.791 | -16.670 | 1.00 |
| βt_hval - TRANSPORT TIME for high value/tonne goods | -3.020 | 0.587 | -5.130 | 0.23 |
| βt_lval - TRANSPORT TIME for low value/tonne goods | -2.470 | 0.443 | -5.580 | 0.19 |
| βpunct_hval - PUNCTUALITY for high value/tonne goods | 5.220 | 1.070 | 4.900 | -0.40 |
| βpunct_lval - PUNCTUALITY for low value/tonne goods | 5.920 | 0.797 | 7.420 | -0.45 |
| βrisk_hval - RISK OF DAMAGE for high value/tonne goods | -50.200 | 4.520 | -11.100 | 3.80 |
| βrisk_lval - RISK OF DAMAGE for low value/tonne goods | -43.200 | 3.130 | -13.800 | 3.27 |
| Null loglikelihood | -1091.014 | | | |
| Cte loglikelihood | -1077.757 | | | |
| Final loglikelihood | -720.876 | | | |
| $\rho^2$ | 0.339 | | | |
| Adjusted $\rho^2$ | 0.332 | | | |

**Fig. 13.24** Relative model estimations, with separate coefficients by value/tonne of transported goods, based on 85 choice experiments

less capital invested in these goods (per tonne), and then the interest costs would be lower.

#### 13.6.2.4 Estimations on 85 Benchmark Shipments with Separate Coefficients by Cargo Value/Tonne (Low Value/Tonne <= 30,000 €/tonne and High Value/Tonne >30,000 €/tonne)

In order to analyse shippers' preferences on the basis of the type of transported goods (goods with low or high value/tonne) the following model was estimated (given the quite small sample size, different transport modes were not considered to get significant results):

$$U_i = ASC_i + \beta_c c_i + \sum_{k=t,p,r} \beta_{k\,highval} x_{ik} d_{highval} + \sum_{k=t,p,r} \beta_{k\,lowval} x_{ik} d_{lowval}$$

where

$$d_{highval} = \begin{cases} 1 & \text{if cargo value / tonne } > 30,000€/\text{tonne}, \\ 0 & \text{otherwise} \end{cases}$$

$$d_{lowval} = \begin{cases} 1 & \text{if cargo value / tonne } \leq 30,000€/\text{tonne}, \\ 0 & \text{otherwise} \end{cases}$$

Estimation results and WTP values are shown in Figs. 13.24 and 13.25.

| ALL BENCHMARK SHIPMENTS - 85 INTERVIEWS | | |
|---|---|---|
| WILLINGNESS TO PAY | €/transport | €/tonne |
| Reduction of 10% in transport time for shipments carrying low value/tonne goods | 14.997 | 1.402 |
| Reduction of 1h in transport time for shipments carrying low value/tonne goods | 1.751 | 0.164 |
| Reduction of 1day in transport time for shipments carrying low value/tonne goods | 42.032 | 3.928 |
| Reduction of 10% in transport time for shipments carrying high value/tonne goods | 6.721 | 17.530 |
| Reduction of 1h in transport time for shipments carrying high value/tonne goods | 0.625 | 1.629 |
| Reduction of 1day in transport time for shipments carrying high value/tonne goods | 14.992 | 39.101 |
| Increase of 10% in the number of shipments/year arriving on time carrying low value/tonne goods | 35.943 | 3.359 |
| Increase of 5 shipments/year arriving on time carrying low value/tonne goods | 6.011 | 0.562 |
| Increase of 10% in the number of shipments/year arriving on time carrying high value/tonne goods | 11.617 | 30.300 |
| Increase of 5 shipments/year arriving on time carrying high value/tonne goods | 1.182 | 3.083 |
| Reduction of 10% in the number of shipments/year suffering damage and/or loss carrying low value/tonne goods | 262.289 | 24.513 |
| Reduction of 5 shipments/year suffering damage and/or loss carrying low value/tonne goods | 43.861 | 4.099 |
| Reduction of 10% in the number of shipments/year suffering damage and/or loss carrying high value/tonne goods | 111.723 | 291.386 |
| Reduction of 5 shipments/year suffering damage and/or loss carrying high value/tonne goods | 11.368 | 29.649 |

**Fig. 13.25** WTP for an improved quality of the average shipment attributes computed on the basis of 85 choice experiments

The trade-off ratios of low value/tonne shipments are similar but always a little bit lower (excluding the punctuality trade-off ratio) than those of high value/tonne shipments. Looking at the WTP values/tonne (€/tonne) it's possible to state that logistics managers shipping higher value/tonne goods are willing to pay more for improvements in the quality of service than shippers transporting low value/tonne goods, as expected.

## 13.7 Conclusions

A conjoint analysis study based on data collected through both a revealed and a stated preference survey was carried out in North West Italy in order to estimate shippers' preferences for freight transport service attributes. The attributes selected to be investigated were: transport cost, transport time, punctuality and risk of damage and loss. Logistics managers of manufacturing firms (food and beverages excluded), in charge of organizing shipments for their companies were interviewed.

The empirical study and the subsequent model estimation provided interesting results, showing that the attributes having more influence (per % change) on the logistics managers' decision making are the risk of damage and loss and the transport cost.

Moreover, shippers evaluate quality attributes such as punctuality and reliability (in terms of avoidance of damages and losses) more important than travel time savings. Transport duration is of limited importance for the interviewees since a shorter transport time only decreases their inventory costs during the shipment, while a higher number of shipments arriving late or suffering damages produces bigger losses such as interruption of the production chain or loss of reliability as a supplier.

Logistics managers indicate a high willingness to pay to improve freight transport service quality, especially to increase their level of reliability and safety. Shippers using intermodal services would pay more than shippers transporting by road to reduce or increase the same quality attributes (both punctuality and avoidance of damages and losses) by the same percentage. Moreover, as expected, logistics managers shipping higher value/tonne goods are willing to pay more for improvements in service quality than shippers transporting low value/tonne goods.

These outcomes clearly show that, from shippers' point of view, modal shift policies should focus more on quality attributes of the mode that is being promoted, such as punctuality and reliability, than on aspects like transport time, especially when considering intermodal transport.

The survey was targeted at a specific market segment and geographical area along Corridor Rotterdam-Genoa (manufacturing firms located in North West Italy), therefore, even if the estimated coefficients and trade-off ratios are similar to those of other studies carried out in different European Countries, the results are not representative for the entire freight market along the Corridor. Specific interviews to manufacturing firms located in other Countries along Corridor 24 would allow to get more specific estimates for the study area.

## References

1. Beuthe M et al (2002) A multi-criteria methodology for stated preferences among freight transport alternatives. In: 7th NECTAR conference: a new millennium. Are things the same? Umea
2. Danielis R, Rotaris L, Marcucci E (2005) Logistics managers stated preferences for freight service attributes. Transp Res Part E Logistics Transp Rev 41:201–215
3. Danielis R, Marcucci E (2006) Trasporto stradale o intermodale ferroviario? I risultati di un'indagine sulla struttura delle preferenze di alcune aziende manifatturiere italiane. Working paper n. 107, Dipartimento di scienze economiche e statistiche, University of Trieste
4. Danielis R (2002) Freight transport demand and stated preference experiments. Milan, Franco Angeli

5. Jong G, Bakker S, Wortelboer-van Donselaar P (2004) New values of time and reliability in freight transport in the Netherlands. In: Proceedings of the European transport conference
6. Jong G (2008) Value of freight travel-time savings, revised and extended chapter for handbooks in transport I. Handbooks of Transport Modelling, Elsevier, Amsterdam
7. Kofteci S, Ergun M, Ay HS (2010) Modeling freight transportation preferences: conjoint analysis for Turkish region. Sci Res Essays 5(15):2016–2021
8. Maggi R, Rudel R (2008) The value of quality attributes in freight transport. Evidence from an SP-Experiment in Switzerland. In: Ben-Akiva M, Meersman H, Van de Voorde E (eds) Recent developments in transport modelling: lessons for the freight sector. Emerald, West Yorkshire
9. Marcucci E (2005) I modelli di scelta discrete per l'analisi dei trasporti. Carrocci, Roma
10. Masiero L, Hensher DA (2010) Analyzing loss aversion and diminishing sensitivity in a freight transport stated choice experiment. Transp Res Part A 44:349–358
11. Masiero L, Maggi R (2010) Accounting for WTP/WTA discrepancy in discrete choice models: discussion of policy implications based on two freight transport stated choice experiments. Quaderno N. 11-03, Faculty of Economics, University of Lugano
12. Orme BK (2005) Getting started with conjoint analysis: strategies for product design and pricing research. Research Publishers, Madison, USA
13. Ortuzar J, Willumsen LG (2001) Modelling transport, 3rd edn. Wiley, Chichester
14. Shinghal N, Fowkes T (2002) Freight mode choice and adaptive stated prefereces. Transp Res Part E 38:367–378
15. Significance, VU University Amsterdam, John Bates Services, in collaboration with TNO, NEA, TNS NIPO and PanelClix (2013) Values of time and reliability in passenger and freight transport in The Netherlands, report for the Ministry of infrastructure and the environment. Significance, The Hague
16. Train KE (2009) Discrete choice methods with simulation, 2nd edn. Cambridge University Press, New York
17. Zotti J, Danielis R (2004) Freight transport demand in the mechanics' sector of Friuli Venezia Giulia: the choice between intermodal and road transport. European Transport/Trasporti Europei 25–26:9–20

Printed by Printforce, the Netherlands